CRITICAL ACCLA

I0044918

"**A Masterfully Written Guide to Understanding War, Strategy, and Leadership in Times of Global Crisis.** Few books arrive at the exact moment when they are needed most, but *Warriors Hate War by Glenn Sturm* is one of them. In a world increasingly shaped by military conflict, geopolitical uncertainty, and shifting alliances, this book serves as an essential guide for leaders who seek to understand the true nature of war—and, more importantly, how to win. At its core, *Warriors Hate War* is a meticulously researched, brilliantly organized, and deeply insightful examination of historical military strategies. Glenn Sturm distills decades of warfare, leadership, and policy decisions into a critical tool for policymakers, military strategists, and business leaders alike.
— John Lyons—Global Entrepreneur | Economic Development & Strategic Support for Ukraine

"Warriors Hate War systematically looks at the past involvement of the United States military in conflicts around the world and examines patterns and outcomes. The book brings much-needed research, analysis, and synthesis and logically proposes a better way not to repeat the dysfunctions of the past. True warriors understand, as no others can, the true cost of war."
—Gray Johnson—Former United States Army Company Commander

"Glenn's visionary business acumen, legal mind, and lifelong passion for military history equip him to uncover unique and valuable perspectives in Warriors Hate War—an insightful read."

—Kenneth M. Ford—CEO of Florida Institute for Human and Machine Cognition.

"I have known Glenn Sturm the power lawyer, entrepreneur, the business leader, and the man with the public company Midas touch. Imagine my surprise when I met Glenn Sturm the astrophotographer. Considering his polymathic expanse of skills, it should have been no surprise to meet Glenn Sturm the military historian but his book, Warriors Hate War— Strategy, politicization, successes, and failures of the American Military encouraged, informed, and amazed me. His thesis of applying the Powel Doctrine to American Military actions as a strategic planning tool for future events is profoundly astute and learning of the Sigma War Games and the extent of their predictive outcomes evokes wonder at why certain actions were ever undertaken considering the accuracy of the predictions. A thoroughly worthy and unique literary contribution in the military history genre."
—Governor David A. Paterson—55th Governor of New York.

"As a military veteran and strategic thinker, Glenn weaves together profound insights with a deep understanding of the human spirit. His contributions inspire those around him to strive for excellence and service. Thanks to his new book, Warriors Hate War, Glenn's impact continues to resonate in meaningful ways as he introduces concepts to inform strategic military thinking. In summary, I believe that this book should make Americans and our leaders think carefully about the standards we should follow before we deploy components of our military."
—Dr. Mark Mykityshyn—Former Chairman, U.S. Army War College Board of Visitors. Dr. Mykityshyn Graduated from the Naval Academy and earned his Ph.D. from Georgia Tech. He served on active duty as an officer in the US Marines as an Aviator and flew with the Navy's flight demonstration team.

WARRIORS HATE WAR

Strategy, politicization, successes, and failures
of the American military

by
GLENN STURM

E&R

PUBLISHERS OF O.G AUTHOR GENIUSES

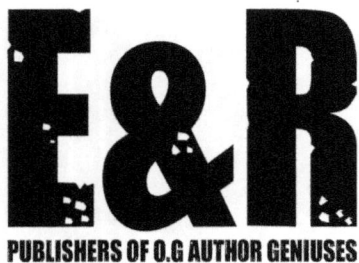

Published by E&R Publishers
New York, NY, USA

An imprint of MillsoCo Publishing, USA
www.EandR.pub

Copyright: © 2025 Glenn Sturm. All rights reserved.

Except for brief quotations in critical articles or reviews, no part of this book may be reproduced in any manner without prior written permission from the publisher. Write to: Permissions, Publisher name, E&R Publishers, New York, NY, USA.

ISBN: 9798990521780 (Hardcover)
ISBN: 9798990521797 (Paperback)
ISBN: 9781966155003 (EBook)
Library of Congress Control Number: 2024945586

DEDICATION

To those like me who wish to leave people, places, things, and the world better than we found it.

ACKNOWLEDGMENTS

Dr. Simon Mills, this book and the three other books that are in various stages of preparation would never have come to fruition without your help, guidance, support, and creativity. You are one of the most talented individuals I have ever met. Thank you, Simon, for your wizardry.

Daniel B. Hodgson, you were my mentor for most of my career. You guided me as I transitioned from a young Army officer to a very young partner and then managing partner of our firm. I would never have arrived where I am without your guidance and friendship.

Steven Philbrick, you were my first company commander in the Army, which is why I became a line company commander as a second lieutenant and a brigade headquarters company commander as a first lieutenant. I always look at you as the officer who set the example for how to lead a military organization.

Col. Earl C. Sturm, Dad, I miss you so much. I loved having a dad who was also my mentor. I remember those phone calls to you when, as a young officer, I was way over my head, and you always answered the phone and provided sage advice. I can't imagine the decisions I would have made without you. I, along with the soldiers I led, continue to be grateful for your help.

TABLE OF CONTENTS

PROLOGUE

What right or authority do I have to comment on the goings-on of the American military over decades of war? I was not a general, a politician, or even a military strategist by profession. But here is the thing: as a lawyer, I spent a lifetime paying attention. Much of that legal career was also spent as an officer in the military on active duty or in the reserves. The rest of the time, I was studying military history, strategy, and process. But let's start before me:

Col Earl Claude Sturm
Memorial Service
2:00 PM
September 2, 2012
Hamilton Mill United Methodist Church

Mom, Marybeth, Barb, Uncle Red, Aunt Edith, Solomon, the rest of our family in attendance, and all our family's friends, thank you for the opportunity to speak at the celebration of our father's life. I am honored to be in your presence.

Dad was my teacher, coach, and mentor. He was one of three people in my adult life to whom I could ask a question and receive an unvarnished answer. He cared enough about me to tell me what I needed to hear, not what I wanted to hear. He was my father and my friend. I truly love him.

Dad lived his life as an Army officer. His perspective was that of a person who had routinely faced life-and-death situations. The fundamental message that my dad taught me is that a leader leads by what he does, not what he says. A leader leads with his own behavior, not by the rules he establishes. He lived his life by setting an example for others to emulate. He knew to lead from the front. He epitomized the Infantry motto: Follow me.

While Dad was successful at the Pentagon, he hated it there. All he ever wanted to do was lead and command.

To quote from the US Army's FM 101.1: Command is the art of decision-making and of leading and motivating soldiers and their organizations into action to accomplish missions at the least expense in manpower and material. Command is vested in an individual who has total responsibility. The essence of command is defined by the commander's competence, intuition, judgment, initiative, and character, and his ability to inspire and gain the trust of his unit. Commanders possess authority and responsibility and are accountable while in command.

Dad pounded two messages into me as a young man. First, a decision by indecision is the worst decision of all. You cannot believe how many times he told me his motto. He stamped it onto me. He also pounded into me the humility and humanity that goes with having the power to make decisions. He told me to identify the problem, get the facts, and make the decision, but be prepared to alter the decision if you get new facts. He reminded me that I had to constantly improve. As a leader, you owe it to your team to constantly improve yourself. Finally, he taught me a simple but oft-forgotten concept that is used in all military affairs: after action reports, if you make a mistake, admit it and publish it so that others can learn. The reason for this process is simple: it is so that neither you nor others will repeat your mistake.

Second, Dad pounded into me the message that if you are late for the "Line of Departure," people die. A lot of the time, I viewed my dad as a simple man. I viewed Dad as a great leader, but not complex. I was completely wrong. It turns out that this simple phrase and most everything Dad communicated was clear, not simple, but the ideas had many levels; they were extremely complex.

In the military, all movements are coordinated. It must be teamwork. If someone misses a deadline, the whole unit can fall into a disorganized mess. A mission can fall apart, and as a result, people can die.

After decades of reflecting on this simple message, his real lesson finally became clear to me while I was writing this eulogy: hitting the "Line of Departure" on time was a metaphor for proving to others that you can depend on you—that you lead by setting the example with your own behavior. Having the reputation of making the "Line of Departure" on time means that others can depend on you—that's what matters! Being dependable.

Dad's real message was that you must be trustworthy. Whether it's your adjacent commanders or the other person in the foxhole with you, that person must trust you. Being on time means that the other commanders can trust you. Being both trustworthy and dependable are two of the foundational requirements for being a leader.

These two characteristics that Dad macerated into me also taught me that to lead I had to be a servant of others. Dad did everything he could to develop me into a servant leader. I hope that I met his standards because Dad's behavior epitomized the concept of being a servant leader.

My Dad taught me that you won't know if you have met the standard until you leave this worldly existence because it is something that you must focus on each day while you are on this planet.

Dad met the standard.

I have been asked to share a few specific memories of Dad, not that anything I say will encapsulate the impact that Dad had on our lives as a teacher and servant leader; but I'll try.

The first is a story that can be summarized by one word: SOUP. It was a requirement in every unit Dad commanded, and by the way, a few others, including me, copied my dad.

Dad assumed command of a Company in combat when he was 19. His first edict was that hot soup would be available 24-7. His troops would always be able to run a few yards, at any time, to get soup. Whether they were cold, hungry, just got back from a patrol, or were getting ready for a mission, there was always hot soup for them.

The simple act of ensuring that the troops always had hot soup proved to the soldiers that the command element was thinking of them and was there to support the troops. As a result of this and other simple acts of kindness, the troops knew he was a commander they could trust.

I later learned that Col. Charles Beckwith adopted the same technique. Col Beckwith is another one of our great commanders, our great military leaders, and the first commander of SFOD-Delta. He also knew you must focus totally on the troops—developing trust and always being there for them. For some reason, this simple concept is one of the reasons we have a great military.

The second story I would like to share with you happened during Dad's first command, a 19-year-old company commander who became a 19-year-old Battalion Commander. Over the years, I have heard different people tell me similar versions of the events of that day. Just thinking about it makes me smile.

After the Battle of the Bulge, Dad was going to give up his command and move to the regimental staff. You see, they didn't know what to do with a now 20-year-old 1st Lieutenant Battalion commander. He couldn't stay as a Battalion Commander he was too junior. He couldn't go back to being a Company commander because he had already been a Battalion Commander.

So, the powers dictated that Dad would move up. The powers decided that he would be the Assistant Regimental S-3, a Major's slot. Before the change of command, there was a "beer blast."

During the party, Dad was doing paperwork at his desk. Several of Dad's soldiers burst into his tent office. These soldiers insisted that Dad come join the party. Dad's Sergeant Major watched the interaction for a few minutes, snickering all the time. The Sergeant Major couldn't stand it any longer and started laughing so hard that he started crying.

The soldiers were annoyed by the Sergeant Major's behavior and asked why he was laughing. The Sergeant Major asked a question. "How old do you have to be to drink?" The soldiers quickly replied, "21." The Sergeant Major then asked the soldiers, "How old is the Old Man?" The Sergeant Major then quickly answered for them, "He ain't there yet."

The Sergeant Major then explained that the CO, our dad, wasn't old enough to drink. He told the soldiers that if Dad went to the party, he would have to enforce the rules. He told them that the "Old Man" knew that the entire organization had earned the beers, but he would not personally violate the rules and would not let violations that he witnessed go uncorrected.

As a result, he decided not to attend his own party because even though he was the "Old Man," the "Old Man" wasn't old enough to drink. He was taking care of his troops; he subordinated his personal enjoyment so that his troops would benefit. That was his responsibility. That is Dad's legacy.

Dad always showed me over the years that he was a man of character, a person who you should trust, and a person for all of us to emulate. His behavior in this case showed what he always taught me. Take care of your troops. That is your first job.

Finally, I want to tell you about what Dad did after he retired from the US Army. Initially, Dad was a little bit lost. He wasn't a leader anymore; he was just another retiree.

Every afternoon, he played tennis with friends. You see, Dad had another great talent; He was a former Texas tennis champion. After a few months, a young man who had been hanging around the tennis courts asked Dad if he would teach him how to play. Dad said yes but had a few rules:

1. Always be on time. 2. Get a note from his teacher every six weeks that he is doing well in school. 3. Show others respect.

After a few weeks, this young man started bringing other young teenagers with him. The number quickly surpassed the number of children that Dad could supervise so he developed a unique aspect of the program, which was the pairing of young tennis players with retirees. During the next few years, Dad also developed a corresponding golf program.

Dad found funding for these programs and all children—no matter what their resources—were able to participate. Estimates vary greatly, but it is believed that over the next 25 years almost 200,000 children participated in the program. Over time, the retirees established three requirements that were absolute, and a few more evolved:

1. Set a good example for others with your behavior. 2. Make sure your teammates have the resources that they need. 3. Be on time, dependable, and trustworthy. Sounds familiar.

As a result of the programs that Dad developed, he received a lot of awards, but the only reward that mattered to him was the satisfaction of helping these children. He determined that children spelled love with the four letters T.I.M.E., and that's what Dad gave them.

The other night, I told my son Daniel that if I was half the man that my dad was, I would consider myself a success. I told Daniel that if I was able to teach him the lessons that Dad taught me and that if Daniel understood them and acted upon them, I would know that I had achieved what my dad would have wanted me to achieve.

Daniel initially didn't understand how I could believe that his achieving half of what his grandpa had achieved was a success. I told him that I hadn't been tested like my dad had. I hadn't had to face the adversity that Dad had faced. As a result, I don't know how I would have reacted to that adversity. I hope that I would have met the challenges, but I don't know.

What we do know is that Dad was tested, and time after time, he always met the challenges. Dad was a tested hero, not a paper one. Again, Dad showed me to lead with my behavior, not my words.

Finally, I want to remind everyone that Dad was an amazing teacher. He taught young soldiers in World War II, Korea, and Vietnam. He taught officers, and the list of his former junior officers who achieved senior rank is endless. He taught thousands of youngsters through his tennis program. Most importantly, however, he taught his family. He taught us to serve, to lead, to care, to admit our mistakes—and most importantly—that what our children need the most is time and a good role model.

Today is not a day to mourn; it is a day to celebrate. Let's celebrate Dad's life, learn from him, remember what he taught us, how he handled difficult situations, the examples he shared with us, how he led from the front, that he walked the walk, and remember to teach others what he taught us.

If we teach Dad's lessons to others and then they continue to teach those lessons to others, then Dad will live forever because he has changed lives because of the lessons that he taught. He will continue to teach, coach, and mentor all of us and continue to do so for generations to come.

May God Bless Colonel Earl Claude Sturm, US Army, Infantry, because God blessed us with this wonderful man for over 88 years.

INTRODUCTION

Why do warriors hate war? Because their bosses don't understand it. What's the mission? How do we win? What is winning? Bush 43 said that winning was when the major combat operations were ceasing. Knowing how to win a war and preparing to win a war if it becomes necessary is part of the solution, but the preparation that prevents or minimizes a war makes a real warrior.

The reason I wanted to write this is to tell the true story. I served as an officer in the United States on active duty and in the reserves for 37 years. As a result of my service, I have seen many things that leaders have done throughout history that are simply unthinkable.

When my first book, *Cancer Set Me Free—turning crisis into calm to survive anything*—became a bestseller, it gave me a platform from which to write. My second book, which will also be published this year, has already sold out all the limited editions and has not been formally announced yet. Its website, syzygies. com, which we launched before the last eclipse, developed a database of approximately 150,000 individuals. With this evolving platform, I wanted to share what I consider to be critically important with as many people as possible.

This story must be told, I am compelled to document it, and as we did post World War II, those people responsible for the unconscionable behavior, incompetence, and dereliction of duty must be brought to justice. If we don't do that, we are not only permitting but encouraging others to do the same again.

As I write this book, two major wars are ongoing. The first is the Russo–Ukrainian War, and the second is the Israeli–Hamas War which, although a cease fire agreement has been reached, this conflict seems far from over. Both wars receive substantial support from the world's major military powers and each of the major powers' allies. In each case, one of the belligerents launched a surprise attack, and the first casualties were civilians.

The atrocities in the Russo–Ukrainian War occurred not only at the beginning of the conflict but continue every day. According to Russian Children's Commissioner Maria Lvova-Belova, as of July 2023, over 700,000 children have been taken/kidnapped since the start of the war[1]. As a result of these atrocities, the International Criminal Court (ICC) in The Hague issued an arrest warrant for Russian President Vladimir Putin and Maria Lvova-Belova.

Putin's attack on the children's hospital on July 8, 2024, reflects his intent to continue the atrocities. Since beginning its invasion, Russia has targeted Ukrainian schools, hospitals, and residential areas, bombed humanitarian corridors, and used prohibited munitions such as cluster bombs. Putin committed similar acts

---

[1] More than 700,000 Ukrainian children taken to Russia. Scan to view.

when he attacked and temporarily conquered Afghanistan. He is a brute and an assassin and must be held accountable.

The Israeli–Hamas War is substantially more complex than the Russo–Ukrainian War. The substantial reason for this was first the deception of the Arabs and the Jewish people. First and repeatedly by the British and second by France. The promises made to the Arabs, combined with the geopolitical agreements that followed, created a sense of betrayal and injustice that still have repercussions in the region today.

The British did not honor the establishment of an independent Arab state for several reasons: the roots of the modern Arab–Israeli conflict lie in the tensions between Zionism and Palestinian nationalism, the latter having risen in response to Zionism toward the end of the 19th century. Territory regarded by the Jewish people as their historical homeland is also considered by the Pan-Arab movement as historically and presently belonging to the Arab Palestinians. Palestine had been under the control of the Ottoman Empire for nearly 400 years until its partitioning in the aftermath of the Great Arab Revolt during World War I.

During the closing years of their empire, the Ottomans began to espouse their Turkish ethnic identity, asserting the primacy of Turks within the empire, leading to discrimination against the Arabs. The promise of liberation from the Ottomans led many Jews and Arabs to support the Allied powers during World War I, leading to the emergence of widespread Arab nationalism. Both Arab nationalism and Zionism had their formulative beginning in Europe. The Zionist Congress was established in Basel in 1897, while the "Arab Club" was established in Paris in 1906.

In the late 19th century, European and Middle Eastern Jewish communities began to increasingly immigrate to Palestine and purchase land from the local Ottoman landlords. The population of the late 19th century in Palestine reached 600,000—mostly

Muslim Arabs, but also significant minorities of Jews, Christians, Druze and some Samaritans, and Bahá'ís. At that time, Jerusalem did not extend beyond the walled area and had a population of only a few tens of thousands. Collective farms, known as kibbutzim, were established, as was the first entirely Jewish city in modern times, Tel Aviv.

During 1915–16, as World War I was underway, the British High Commissioner in Egypt, Sir Henry McMahon, secretly corresponded with Husayn ibn 'Ali, the patriarch of the Hashemite family and Ottoman governor of Mecca and Medina.

McMahon convinced Husayn to lead an Arab revolt against the Ottoman Empire, which was aligned with Germany against Britain and France in the war.

McMahon promised that if the Arabs supported Britain in the war, the British government would support establishing an independent Arab state under Hashemite rule in the Arab provinces of the Ottoman Empire, including Palestine. The Arab revolt, led by T. E. Lawrence—"Lawrence of Arabia"—and Husayn's son Faysal, was successful in defeating the Ottomans, and Britain took control over much of this area.

You may be wondering why I would include the modern Israeli–Arab conflicts in the manifest of US military actions. Well, because America's involvement in the modern Arab–Israeli conflict has been extensive and multifaceted, reflecting a mix of political, military, and economic interests. We will explore these areas more in-depth later in the book.

The Israeli–Hamas War is substantially more complex than the Russo–Ukrainian War. A substantial reason for this war was a deception brought upon both of the parties by one of the major powers toward the end of the World War I. During the period from 1915 to 1916, Sir Henry McMahon, the British High Commissioner for Egypt, negotiated through correspondence

4

with Sharif Hussein, who was a member of the Hashemite family, which claimed descent from the Prophet Muhammad. These letters, known as the McMahon–Hussein Correspondence, promised support for Arab independence if the Arabs revolted against the Ottoman Empire.

From 1916 to 1918, Sharif Hussein's sons, with the support of T.E. Lawrence (AKA Lawrence of Arabia), led a revolt against the Ottoman Turks. The revolt was successful in weakening the Ottoman's control over the Arab regions. However, during the early part of the revolt, another secret agreement between Britain and France negotiated an agreement whose purpose was for France and Britain to split control over the Arab regions between themselves. This agreement was not disclosed until after the war ended and clearly contradicted the agreement with the Arabs.

To compound the deceit, in 1917, Britain issued the Balfour Declaration. The key passage stated:

> His Majesty's Government view with favor the establishment in Palestine of a national home for the Jewish people and will use their best endeavors to facilitate the achievement of this object, it being clearly understood that nothing shall be done which may prejudice the civil and religious rights of existing non-Jewish communities in Palestine. Many individuals believe that the declaration also assumed that there would be an independent Arab state[2].

The promises made to the Arabs, combined with the agreements between Britain and France to divide the "Arab" lands that followed and the creation of Israel, created a sense of betrayal and

---

[2]Balfour Declaration 1917. Scan code to view.

injustice that lasts to this day. Now Trump has proposed a Middle East riviera with resorts and casinos. I'll withhold commentry.

One of the things we should have learned from what happened before World War II is that appeasement only empowers evil people such as Hitler and Putin. That's the reason I believe that this book needs to be written and the horrific individuals who are responsible for the atrocities taken before a court and appropriately held to account. After World War II, it took almost 70 years before some of the guilty saw justice. Let's hope we can do better this time.

It has taken the United States two years to begin assisting Ukraine in the Russo–Ukrainian War, and depending on elections and administrations, it may not be long before that assistance sees an end.

I may be a foolish idealist, but there's a reason I have survived cancer for over 16 years. I will not go to my deathbed for many years, and I won't go without doing all I can to hold these people accountable.

We will explore many aspects of military action, including battles that should not have been fought, battles that should have been fought but were not, small insignificant decisions with massive cascading ramifications, and most importantly, what I mean when I say Warriors Hate War.

There is an important phrase that I use a lot, and it is often misunderstood. The phrase is "Only you can stop you." This wonderfully simple phrase is one of the most complex aspects of being an Army leader. Most leaders think it simply means commit to accomplish the mission, never stop, never let up because only you can stop you from accomplishing the mission. But it's a lot more complex than that. There are times when you will be on a mission without the ability to communicate with higher headquarters. That could result in an obvious mission change, or a contingency occurs that changes what you need to do. Since you are the decision–maker, you need to stop, reassess, and once you have decided to do so, you will be on your way to success. In this case, you must stop you. Applying this methodology to many of the incidences we will explore in this book would have produced different outcomes in many of the scenarios and situations.

Finally, there are three other significant issues facing the US military. First, there is an article published by the US Army War College (USAWC) Press entitled *Lying to Ourselves: Dishonesty in the Army Profession*,[3] which addresses, not surprisingly, dishonesty in the Army profession. It is a thoughtful article that underscores the next two issues. The second issue is the politicization of the Officer Corps[4]. The third article is *Anatomy of Failure: An Analysis of Why America Keeps Losing Wars*[5].

We will review several standards that will predict the outcome of future military conflicts, primarily by reviewing the doctrines of our country's civilian and military leaders since 1947.

I believe that we should look at how our major theater commanders in the European Theater of Operations and the Pacific Theater of Operations during World War II conducted themselves and the important decisions they made upon the conclusion of

---

[3]Lying to Ourselves: Dishonesty in the Army Press: Dishonesty in the Army Profession Leonard Wong. Dr. SSI Stephen J. Gerras Dr.; US Army War College USAWC Press, 2-1-2015.

[4]The Increasingly Dangerous Politicization of the U.S. Military. David Barno and Nora Bensahel, War on The Rocks.

[5]Anatomy of Failure Adm. James Stavridis.

World War II. Some of those decisions laid the groundwork for the economic and key military relationships we have today.

We will start by identifying all the major conflicts that the United States has been directly or indirectly involved in since the end of World War II. We will also compare the one spectacular victory and the doctrine that predicated that conflict's success with the doctrines that were in place when we regularly lost.

We will also examine the ethics that permitted us to run the School of the Americas, where we trained foreign officers. Many of its graduates have been involved in human rights violations and atrocities in their countries.

Our hope is that this book will provide American citizens with a straightforward set of questions they can ask our leaders to justify proposed military actions and, as a result, prevent future poorly analyzed use of the US military in unjustified military conflicts.

US MILITARY ACTIONS SINCE WORLD WAR II

The following is an overview of each of the major US military actions since World War II, in which the US military was directly or indirectly involved. I have included several of Israel's conflicts because of our financial and military aid to Israel. Furthermore, there is a significant overlap between our two countries' simultaneous involvement in major conflicts at the same time.

1940s

1. **Greek Civil War (1947–1949):** After World War II, a civil war broke out in Greece between government forces (supported by the United States and Britain) and communist rebels (supported by the Soviet Union and Yugoslavia). The United States provided military and economic aid under the Truman Doctrine, fearing the spread of communism in Europe.

2. **Berlin Blockade and Airlift (1948–1949):** In response to Soviet attempts to blockade Berlin and force the allies out of the city, the United States and its allies launched a massive airlift to supply West Berlin with food and fuel. This became a major Cold War confrontation and lasted nearly a year.

3. **Israel's War of Independence (1947–1949) and the Palestinian Nakba** (Arabic for "catastrophe") are two deeply intertwined and significant events in the history of the Israeli–Palestinian conflict.

> **Israel's War of Independence (1947–1949):** The Israel War of Independence, also known as the 1948 Arab Israeli War, began following the United Nations (UN) resolution on November 29, 1947, which proposed the partition of British Mandate Palestine into two states— one Jewish and one Arab—while maintaining Jerusalem as an international city. The Jewish leadership in Palestine accepted the UN plan, but the Arab leadership rejected it.

> **Phase 1 (Civil War, November 1947–May 1948):** After the UN partition plan was announced, violence erupted between Jewish and Arab communities in Palestine. This phase is sometimes referred to as the Palestinian Civil War. It included skirmishes, attacks, and counterattacks between Jewish and Arab militias, leading to increased tensions and displacements of populations.

> **Phase 2 (May 1948–March 1949):** On May 14, 1948, the State of Israel declared its independence. The next day, May 15, armies from surrounding Arab nations, including Egypt, Transjordan (Jordan), Syria, Lebanon, and Iraq, invaded the newly declared state. The war then

became an interstate conflict between the nascent state of Israel and these Arab nations.

The fighting was intense and lasted until early 1949, with Israel managing to repel the Arab armies and even expanding beyond the borders proposed in the UN partition plan.

By 1949, several armistice agreements were signed, effectively ending the war. Israel retained the areas allocated to it under the UN partition plan as well as additional territories, including the western part of Jerusalem.

Palestinian Nakba (1948): The term "Nakba" refers to the mass displacement of Palestinians that took place during the 1948 war. An estimated 700,000–750,000 Palestinian Arabs were expelled or fled from their homes during the conflict, leading to the creation of a large Palestinian refugee population.

The Nakba is viewed as a direct consequence of the Israeli War of Independence as Jewish forces captured Arab towns and villages, many of which were depopulated and destroyed.

- Some Palestinians fled due to fear of violence or after being ordered to evacuate by local Arab leaders or Arab armies.
- Others were expelled by Jewish forces during and after battles. Incidents like the Deir Yassin Massacre in April 1948, where over 100 Palestinian villagers were killed by Jewish paramilitary groups, fueled fear and contributed to the exodus.

- The Nakba continues to be a central issue in the Israeli–Palestinian conflict. Palestinian refugees and their descendants have maintained the demand for the "right of return" to their former homes, which remains a deeply contested issue in peace negotiations.

Outcomes and Legacy

For Israel: The war resulted in the establishment of the State of Israel as a sovereign nation. The war is seen by Israelis as a struggle for survival and national liberation, leading to the successful creation of a Jewish state in the aftermath of the Holocaust.

For Palestinians: The Nakba is seen as a national tragedy, marking the loss of their homeland and the beginning of the Palestinian refugee crisis. Many Palestinians and their descendants still live in refugee camps or in exile, with the hope of one day returning to their ancestral homes.

Impact on the Israeli–Palestinian Conflict: The war and Nakba cemented deep divisions between Jews and Arabs in the region. The outcome of the war and the subsequent displacement of Palestinians created long-lasting grievances.

The refugee issue, along with the boundaries established by the armistice lines (often referred to as the Green Line), set the stage for ongoing territorial disputes, future wars (such as the 1967 Six-Day War), and numerous attempts at peace negotiations, which remain unresolved to this day.

In essence, the Israel War of Independence and the Palestinian Nakba are pivotal events that shaped the modern

Middle East and the ongoing conflict between Israelis and Palestinians, with both sides having deeply contrasting narratives and experiences of these events.

1950s

4. **Palestinian Fedayeen Insurgency (1950s–1960s):** Palestinian attacks and reprisal operations carried out by the IDF. These actions were in response to constant Palestinian fedayeen incursions during which Arab guerrillas infiltrated from Syria, Egypt, and Jordan into Israel to carry out attacks against Israeli civilians and soldiers.

5. **Korean War (1950–1953):** The Korean War began when North Korea, supported by China and the Soviet Union, invaded South Korea. The United States, as part of a UN coalition, intervened on behalf of South Korea. The war ended in an armistice, but no formal peace treaty was signed until a few years ago, leaving Korea divided.

6. **First Indochina War (1950–1954):** The United States provided military aid to French colonial forces fighting against the Viet Minh, the communist-led independence movement in Vietnam. After the French defeat at Dien Bien Phu, the war ended with the Geneva Accords, dividing Vietnam.

7. **Iranian Coup (1953):** The CIA orchestrated a coup against Iran's democratically elected prime minister, Mohammad Mossadegh, after he nationalized the oil industry. The United States reinstalled the Shah, Mohammad Reza Pahlavi, who ruled as a pro-Western autocrat until the 1979 Iranian Revolution.

8. **Guatemalan Coup (1954):** The CIA supported a coup to overthrow Guatemala's democratically elected president,

Jacobo Árbenz, who had implemented land reforms that threatened US business interests. The coup installed a military dictatorship, leading to decades of civil unrest.

9. **Suez Crisis (October 1956):** A military attack on Egypt by Britain, France, and Israel, beginning on October 29, 1956, with the intention to occupy and to take over the Sinani Peninsula and the Suez Canal from Egypt.

10. **Lebanon Crisis (1958):** US Marines landed in Lebanon to stabilize the pro-Western government during a civil war triggered by regional tensions and concerns about rising communist influence in the Middle East.

11. **Taiwan Straits Crises (1954–1955, 1958):** The United States supported Taiwan during two crises with Communist China, in which the Chinese military shelled islands controlled by Taiwan. The United States sent naval forces to protect Taiwan and deter further aggression.

1960s

12. **Bay of Pigs Invasion (1961):** A failed CIA-backed invasion of Cuba by Cuban exiles aimed at overthrowing Fidel Castro's communist government. The failure was a major embarrassment for the Kennedy administration.

13. **Cuban Missile Crisis (1962):** A 13-day confrontation between the United States and the Soviet Union over Soviet ballistic missiles deployed in Cuba. The crisis brought the two superpowers to the brink of nuclear war before a negotiated settlement was reached, with the Soviets agreeing to remove the missiles.

14. **Vietnam War (1955–1975):** The United States escalated its involvement in Vietnam to prevent the spread of communism in Southeast Asia. US military forces supported the

South Vietnamese government against the communist Viet Cong and North Vietnamese forces. The war ended with the fall of Saigon in 1975, leading to the reunification of Vietnam under communist control.

15. **Dominican Republic Intervention (1965):** During a civil war in the Dominican Republic, the United States sent troops to prevent a communist takeover and protect American citizens. The intervention resulted in the installation of a pro-American government.

16. **Laos (Secret War) (1964–1973):** The CIA conducted covert operations in Laos, supporting anticommunist forces against the Pathet Lao and North Vietnamese troops. The United States also carried out extensive bombing campaigns, targeting communist supply lines and bases.

17. **Six-Day War (June 1967):** The Six-Day War took place from June 5 to June 10, 1967, and was fought between Israel and the neighboring Arab states of Egypt, Jordan, and Syria. It is one of the most significant conflicts in the Arab Israeli conflict and reshaped the geopolitics of the Middle East.

    Israel's victory in the Six-Day War resulted in the capture of significant territories from its Arab neighbors, fundamentally altering the map of the region:

    - Sinai Peninsula and Gaza Strip: captured from Egypt.
    - West Bank, including East Jerusalem: captured from Jordan.
    - Golan Heights: captured from Syria.

Israel's territorial gains gave it control over areas with significant Palestinian populations, which created new political and military challenges for the region. The capture of East Jerusalem had profound religious and cultural significance for Jews, Muslims, and Christians.

18. **War of Attrition (1967–1970):** A limited war fought between the Israeli military and forces of the Egyptian Republic, the USSR, Jordan, Syria, and the Palestine Liberation Organization. It was initiated by the Egyptians as a way of recapturing the Sinai Peninsula from the Israelis, who had been in control of the territory since the mid-1967 Six-Day War.

19. **Cambodia (Secret War) (1969–1973):** US forces, in coordination with South Vietnam, conducted bombing campaigns and ground operations in Cambodia to destroy Viet Cong and North Vietnamese sanctuaries. This escalation contributed to the destabilization of Cambodia and the rise of the Khmer Rouge.

1970s

20. **Invasion of Cambodia (1970):** US and South Vietnamese forces launched a joint invasion of Cambodia to eliminate Viet Cong bases. The incursion sparked widespread protests in the United States, contributing to the growing anti-war movement.

21. **Palestinian Insurgency in South Lebanon (1971–1982):** The Palestine Liberation Organization relocated to South Lebanon from Jordan, staged attacks on the Galilee, and used South Lebanon as a base for international operations. In 1978, Israel launched the first Israeli large-scale invasion of Lebanon, which was carried out by the Israel Defense Forces to expel PLO forces from the territory. Continuing ground and rocket attacks and Israeli retaliations eventually escalated into the 1982 war.

22. **Yom Kippur War (1973):** Fought from October 6 to 26, 1973, by a coalition of Arab states led by Egypt and Syria against Israel as a way of recapturing part of the territories

that they lost to the Israelis back in the Six-Day War. The war began with a surprise joint attack by Egypt and Syria on the Jewish holiday of Yom Kippur. Egypt and Syria crossed the ceasefire lines in the Sinai and the Golan Heights, respectively. Eventually, Arab forces were defeated by Israel, and there were no significant territorial changes.

23. **Mayaguez Incident (1975):** Following the capture of the American merchant ship SS Mayaguez by the Khmer Rouge, US forces launched a rescue mission to recover the ship and its crew. The incident resulted in the deaths of 41 US military personnel, but the crew was safely rescued.

24. **Angolan Civil War (1975–1976):** The United States provided covert support to anticommunist factions in Angola during the civil war that erupted following the country's independence from Portugal. The conflict drew in both the United States and the Soviet Union as part of the Cold War.

1980s

25. **Operation Eagle Claw (1980):** A failed US military mission to rescue hostages held at the American embassy in Tehran during the Iranian Hostage Crisis. The operation ended in disaster when helicopters collided in the Iranian desert, resulting in the deaths of eight US servicemen.

26. **El Salvador Civil War (1980–1992):** The United States provided military aid and training to the Salvadoran government during its civil war against leftist guerrilla forces. The war was part of the broader Cold War struggle between the United States and Soviet-backed communist movements in Latin America.

27. **Lebanon War (1982):** The Israel Defense Forces invaded Southern Lebanon to expel the PLO from the territory on

June 6, 1982. The Cabinet of Israel ordered the invasion as a response to the assassination attempt against Israel's ambassador to the United Kingdom, Shlomo Argov, by the ABU Nidal Organization and due to the constant terror attacks on northern Israel made by the Palestinian guerrilla organizations that resided in Lebanon. The war resulted in the expulsion of the PLO from Lebanon and created an Israeli occupation in southern Lebanon.

28. **Lebanon Intervention (1982–1984):** US Marines were deployed as part of a multinational peacekeeping force during the Lebanese Civil War. In 1983, a truck bomb struck a Marine barracks in Beirut, killing 241 American servicemen, leading to the eventual withdrawal of US forces.

29. **Invasion of Grenada (Operation Urgent Fury) (1983):** US forces invaded the Caribbean Island of Grenada to overthrow a Marxist government and restore order following a coup. The invasion was justified as a rescue mission for American medical students on the island but was widely criticized internationally.

30. **Invasion of Panama (Operation Just Cause) (1989–1990):** US forces invaded Panama to depose its military dictator, Manuel Noriega, who had been involved in drug trafficking. The operation was swift, and Noriega was captured and brought to the United States for trial.

31. **South Lebanon Conflict (1985–2000):** Nearly 15 years of warfare between the Israel Defense Forces and its Lebanese Christian proxy militias against Lebanese Muslim guerrilla, led by Iranian-backed Hezbollah. Within what was defined by the Israelis as the Security Zone in South Lebanon.

32. **First Intifada (1987–1993):** The first large-scale uprising against Israel in the West Bank and the Gaza Strip.

1990s

33. **Gulf War (Operation Desert Storm) (1990–1991):** After Iraq invaded Kuwait, the United States led a coalition of countries in a large-scale military operation to expel Iraqi forces. The war ended with a decisive coalition victory but left Saddam Hussein in power. General Colin Powell was the chairman of the Joint Chiefs of Staff during this war. He employed the Powell Doctrine, and the results speak for themselves.

34. **Somalia Intervention (Operation Restore Hope) (1992–1995):** The United States led a humanitarian mission in Somalia to deliver food and aid during a famine and civil war. The mission turned into a military engagement, culminating in the 1993 Battle of Mogadishu (depicted in the film *Black Hawk Down*), in which 18 US soldiers were killed.

35. **Bosnian War (1992–1995):** The United States, as part of NATO, conducted airstrikes against Bosnian Serb forces committing ethnic cleansing against Bosnian Muslims and Croats. The conflict ended with the Dayton Accords, which brought peace to Bosnia.

36. **Haiti Intervention (Operation Uphold Democracy) (1994):** US forces intervened in Haiti to restore the democratically elected president, Jean-Bertrand Aristide, who had been ousted in a military coup.

37. **Kosovo War (Operation Allied Force) (1999):** The United States participated in NATO airstrikes against Yugoslav forces to stop the ethnic cleansing of Albanians in Kosovo. The conflict ended with a peace agreement and the eventual withdrawal of Yugoslav forces.

2000s

38. **Second Intifada (2000–2005):** The second Palestinian uprising, a period of intensified violence, which began in late September 2000.

39. **War in Afghanistan** (Operation Enduring Freedom) **(2001–2021):** Following the September 11 attacks, the United States invaded Afghanistan to dismantle al-Qaeda and remove the Taliban from power. The war became the longest in US history, with a NATO-led coalition conducting counterinsurgency operations and attempts to stabilize the country.

40. **The Philippines (Operation Enduring Freedom—the Philippines) (2002–2015):** The United States provided counterterrorism training and support to Philippine forces in their fight against Islamist militant groups such as Abu Sayyaf and the Moro Islamic Liberation Front. Sec. Rumsfeld rejected Gen. Shinseki's recommendations along with the Powell Doctrine, and these decisions—along with rejecting Gen. Marshall and MacArthur's doctrines—resulted in a multidecade-long war.

41. **Iraq War (Operation Iraqi Freedom) (2003–2011):** The United States led an invasion of Iraq, citing the need to eliminate weapons of mass destruction (WMDs) and topple Saddam Hussein's regime. The war led to a prolonged occupation, insurgency, and eventually, the rise of ISIS.

42. **Pakistan (Drone Strikes)** (2004–Present): The United States has conducted drone strikes in Pakistan's tribal areas, targeting al-Qaeda, Taliban, and other militant groups. These strikes have been controversial due to civilian casualties.

43. **Lebanon War (Summer 2006):** A military operation that began in response to the abduction of two Israeli reserve

soldiers by Hezbollah. The operation gradually strength-ened to become a wider confrontation. The principal par-ticipants were Hezbollah paramilitary forces and the Israeli military. The war resulted in a stalemate.

44. **Somalia (Counterterrorism Operations)** (2007–Present): The United States has conducted airstrikes, drone strikes, and special operations in Somalia against the al-Shabaab militant group, which is linked to al-Qaeda.

45. **Gaza War or Operation Cast Lead (December 2008–January 2009):** A 3-week armed conflict between Israel and Hamas during the winter of 2008–2009. In an escalation of the ongoing Israeli–Palestinian conflict, Israel responded to ongoing rocket fire from the Gaza Strip with military force in an action titled "Operation Cast Lead." Israel opened the attack with a surprise air strike on December 27, 2008. Israel's stated aim was to stop such rocket fire from, and the import of arms into Gaza. Israeli forces attacked military and civilian targets, police stations, and government buildings in the opening assault. Israel declared an end to the conflict on January 18 and completed its withdrawal on January 21, 2009.

46. **Libya Intervention (Operation Odyssey Dawn) (2011):** The United States participated in a NATO-led military intervention in Libya to support rebels fighting against Muammar Gaddafi's regime. The intervention helped topple Gaddafi, but Libya has remained unstable ever since.

47. **Israeli Operation in the Gaza Strip or Operation Pillar of Defense (November 2012):** Military offensive on the Gaza Strip.

48. **Gaza War or Operation Protective Edge (July–August 2014):** A military offensive on the Gaza Strip as a response

to the collapse of American-sponsored peace talks, attempts by rival Palestinian factions to form a coalition government, the kidnapping and murder of three Israeli teenagers, the subsequent kidnapping and murder of a Palestinian teenager, and increased rocket attacks on Israel by Hamas militants.

49. **Operation Inherent Resolve (2014–2021):** The United States led a coalition to combat ISIS in Iraq and Syria after the group seized large swathes of territory. The operation involved airstrikes, special forces, and support for local ground forces.

50. **Yemen (Counterterrorism Operations)** (2002–Present): The United States has been involved in counterterrorism operations in Yemen, primarily through drone strikes targeting al-Qaeda in the Arabian Peninsula (AQAP) and later, ISIS affiliates.

2020s

51. **Global Drone Campaign (Ongoing):** The United States continues to conduct drone strikes in various countries, including Yemen, Somalia, and Pakistan, as part of its global counterterrorism efforts.

52. **Afghanistan Withdrawal (2021):** After nearly two decades of involvement, the United States completed its withdrawal from Afghanistan in August 2021, leading to the rapid collapse of the Afghan government and the return of the Taliban to power.

53. **Israel–Palestine Crisis or Operation Guardian of the Walls (May 2021):** Riots between Jews and Arabs in Israeli cities. Hamas fired rockets into Israel, with Iron Dome

intercepting the most dangerous projectiles. Israel began airstrikes in Gaza.

54. **Syria (Ongoing Conflict)**: The United States continues to maintain a limited military presence in Syria, primarily to combat remnants of ISIS and prevent the re-emergence of terrorist groups.

55. **Somalia (Ongoing Conflict)**: US forces continue limited counterterrorism operations against al-Shabaab, an al-Qaeda affiliate, through drone strikes and special operations in coordination with Somali forces.

56. **Israel–Hamas War (October 2023–Present)**: Operation Swords of Iron began on October 7, 2023, when Hamas, the Islamist militant group that controls the Gaza Strip, launched an unprecedented surprise attack on Israel. The conflict rapidly escalated into one of the deadliest confrontations between Israel and Hamas in recent history. As of the time of this writing, the Israel–Hamas War is ongoing, with no clear end in sight. Diplomatic efforts for a ceasefire have so far been unsuccessful, and both sides remain locked in a deadly and destructive confrontation. The conflict continues to have profound consequences for both Israelis and Palestinians, as well as for the broader geopolitical dynamics of the Middle East. This war marks the most significant escalation of violence between Israel and Hamas since the 2014 Gaza War, and it has deepened the already entrenched divisions between the two sides. The conflict has also intensified the debate over long-term solutions to the Israeli–Palestinian conflict, including the future of Gaza, the status of Palestinian refugees, and the broader peace process.

Since World War II, the United States has been involved in a near-continuous series of military actions, including wars, conflicts, and smaller-scale operations or interventions. The United States has had very few periods of complete peace without military actions of any kind.

To summarize the comprehensive list above and show the continuity of America's military activity, the timeline looks something like the following:

1945-1950: Post-World War II, the United States was involved in the occupation of Germany, Japan, and Italy, as well as in the Greek Civil War (1946-1949) and the Chinese Civil War (1945-1949). The United States also engaged in actions related to the emerging Cold War, although no large-scale wars occurred until the Korean War began in 1950.

1950-1953: The Korean War was a conflict between North Korea, backed by China and the Soviet Union, and South Korea, supported by the United States and other UN forces. It began when North Korean forces, aiming to unify the Korean Peninsula under communism, invaded South Korea. The war saw fierce fighting, including US-led UN forces pushing the North Koreans back, followed by Chinese intervention that led to a prolonged stalemate. After three years of intense combat, an armistice was signed in 1953, restoring the division at the 38th parallel but leaving North and South Korea technically still at war. The conflict resulted in significant casualties and heightened Cold War tensions. On January 12, 1950, US Secretary of State Dean Acheson delivered a famous speech at the National Press Club in which he failed to include South Korea in America's defense perimeter in the Pacific.

Gen. Dwight Eisenhower, campaigning in the 1952 presidential election, charged that Acheson's omission "gave the green light" to a North Korean invasion because it convinced the Communists that America would not defend the south.

Historians and military analysts would debate the charge's merits, but a public consensus emerged that the Truman administration had bungled by signaling North Korea, China, and the Soviet Union that the United States considered South Korea to be expendable.

1953–1964: During this period, the United States was involved in various Cold War conflicts, including the support of the French in Vietnam, covert operations in Iran and Guatemala, and the beginning of deeper involvement in Vietnam (which escalated after 1964).

1964–1973: Vietnam War, along with other Cold War-related conflicts and interventions, including the Dominican Republic (1965) and actions in Laos and Cambodia.

1973–1980: Post-Vietnam, the United States was involved in smaller operations such as those in Angola, Zaire, and Lebanon, and there was significant military presence and action related to the Cold War, but there was no large-scale war.

1980s: The United States engaged in several military actions, including the Iran hostage rescue mission (1980), intervention in Lebanon (1982–1984), invasion of Grenada (1983), bombing of Libya (1986), and support for anticommunist insurgencies in Latin America and Afghanistan.

1990–1991: Gulf War.

1990s: Post-Gulf War, US forces were involved in enforcing no-fly zones in Iraq, humanitarian interventions in Somalia (1992–1995), intervention in Haiti (1994), and NATO operations in Bosnia (1995) and Kosovo (1999).

2001–present: After 9/11, the United States has been involved in the War on Terror, including the wars in Afghanistan and Iraq, as well as various other military operations around the world.

Conclusions

Given this history, the periods of time when the United States was not involved in any military action, including smaller-scale operations, are minimal. Most historians and analysts would argue that the United States has been continuously involved in some form of military action or intervention since World War II, even if there were brief lulls between larger conflicts.

Thus, virtually no time since World War II has the United States not been involved in some type of military action. For this reason, I wanted to explore military strategy, war crimes, government intervention, military politicization, mistakes, accidents, and what we could do better.

An examination of three major conflicts, the different strategic approaches to the conflicts, and the results of the conflicts should be enlightening. The three we should examine are the Vietnam War, the First Gulf War (the impact of the Powell Doctrine), and the Second Gulf War (the impact of Secretary Rumsfeld's Doctrine). Specifically, Rumsfeld's rejection of General Shinseki's request for a materially larger force.

Again, my wish for this book is for it to serve as a reference for military framework and conduct to improve decision-making and further fortify military strategy for our future. What questions should be asked and answered before we deploy our military in a military action?

THE DOCTRINES

Military doctrine is a set of principles that guide how military forces prepare for and conduct warfare. It includes strategies, tactics, and operational procedures and is developed based on evidence and the best way to do things. A US presidential doctrine comprises the key goals, attitudes, or stances for US foreign affairs outlined by a president. In this book, we will explore some important presidential and military doctrines and use them as lenses to evaluate alternate outcomes.

The Powell Doctrine—The One That Matters— The Bush 41 Doctrine

George Bush, the 41st president of the United States, focused primarily on foreign policy during his time in the White House. During his presidency, Germany was reunifying, the Soviet Union was collapsing, and the Cold War was coming to an end. He improved US–Soviet relations, notably through a meeting with Soviet leader Mikhail Gorbachev, where they signed the Strategic Arms Reduction Treaty in July 1991.

Bush also authorized military operations in Panama and the Persian Gulf. In December 1989, the United States invaded Panama to overthrow dictator Manuel Noriega as president. Noriega was in control of the Panama Canal and was simultaneously running a major drug tariffing operation. Panama had a large population of expat Americans, and he presented a significant threat to American citizens living in Panama.

In October 1989, Bush appointed General Powell as chairman of the Joint Chiefs of Staff. He was 52 years old, which made him one of the youngest chairmen in history. General Powell is regarded as one of the most effective senior officers in American history.

In August 1990, 10 months after General Powell's appointment as chairman, Iraqi leader Saddam Hussein invaded Kuwait and threatened Saudi Arabia. This effectively threatened the free flow of a significant portion of the world's oil supply.

During the buildup to the US response to the Invasion, the United States utilized the Powell Doctrine to prepare for the war. Powell's Doctrine has eight significant components that must be answered in the affirmative before we deploy the US military in a war. The questions and the answers follow:

1. Is a vital national security threatened?

Yes, If Iraq gained permanent control of Kuwait's oil fields Iraq, a middle eastern country would have taken most of the world's oil supply.

2. Do we have a clear attainable objective?

Yes, to neutralize Iraq's very large military organization and return Kuwait's lawful government.

3. Have the risks and costs been fully and frankly analyzed?	Yes, the United States made a profit of at least $48.2 billion fighting the First Gulf War. The primary reason for this was the broad support of international allies and Kuwait's lawful government.
4. Have all other nonviolent policy means been fully exhausted?	Yes, the UN posted several resolutions, including 660, 662, 663, 665, 666, 667, 669, 670, 674, and 677, demanding that Iraq withdraws from Kuwait. On November 29, 1990, the Security Council passed Resolution 678 under the guidance of Canada, the USSR, the United Kingdom, and the United States, which gave Iraq until January 15, 1991, to withdraw from Kuwait, and empowered states to use "all necessary means" to force Iraq out of Kuwait after the deadline. Resolution 678 provided legal authorization for UN members to "use all necessary means" to remove Iraq from Kuwait.
5. Is there a plausible exit strategy to avoid endless entanglement?	Yes, the United States and its allies used the Second portion of the Powell Doctrine that is applied when we actually engage in war: "When a nation is engaging in war, every resource and tool should be used to achieve decisive force against the enemy, minimizing casualties and ending the conflict quickly by forcing the weaker force to capitulate." As a result of the strategy, the ground war lasted less than 100 hours and Iraq capitulated, and we avoided an endless entanglement.

6. Have the consequences of our action been fully considered?	Yes, the consequences of the coalition acting or not acting were considered by the United States and approximately 40 independent countries.
7. Is the action supported by the American people?	Yes, Bush 41 approval rating soared upon the start of the First Gulf War and went higher after the victory.
8. Do we have genuine broad international support?	Yes, over 39 countries supported the First Gulf War.

As demonstrated above, Powell has also asserted that "when a nation is engaging in war, every resource and tool should be used to achieve decisive force against the enemy, minimizing casualties and ending the conflict quickly by forcing the weaker force to capitulate."

Utilizing the Powell Doctrine, President Bush organized a coalition of over 30 countries. This coalition launched a US-led air assault against Iraq in January 1991. After five weeks of air strikes and less than 100 hours of ground combat, Operation Desert Storm concluded with Iraq's defeat and Kuwait's liberation. General Powell's Doctrine helped ensure the coalition suffered minimal casualties while defeating one of the world's largest armies. Some experts say that this is the only military operation since World War II that the United States and its allies have won.

As you will see—from Truman's containment of the communist expansion strategy to Reagan's support of anticommunist guerrillas and resistance movements in Africa, Asia, and Latin America—US foreign policy evolved through different doctrines to respond to shifting opposition strategies and threats. These changes in US tactics were all centered on the US strategy of containing communist expansion and diplomacy during the Cold War. None of them succeeded in terminating our opposition's strategy or tactics.

The reason we started with the Powell Doctrine is that it is the only doctrine that has proved to be very successful. The Powell Doctrine was developed and finalized during a time when the United States was engaged in war. At that time, General Powell used every resource and tool available to apply decisive force against the enemy, minimizing casualties and ending the conflict quickly by forcing the weaker force to capitulate. Furthermore, the Powell Doctrine was successful, while the remaining doctrines were either unsuccessful, failed, or continued the status quo.

The Truman Doctrine

President Harry S. Truman established that the United States would provide political, military, and economic assistance to all democratic nations under threat from external or internal authoritarian forces. The Truman Doctrine effectively reoriented US foreign policy, away from its usual stance of withdrawal from regional conflicts not directly involving the United States to one of possible intervention in faraway conflicts. One of the first uses of the doctrine was to provide troops and advisors to the government of Greece, which was involved in an action against an insurgency backed by the Soviet Union.

The Eisenhower Doctrine

Eisenhower's Doctrine was a continuation of the Truman Doctrine of preventing any extension of Soviet or communist countries in the Middle East. In the "Special Message to the Congress on the Situation in the Middle East," any Middle Eastern country could request that the United States provide military or economic aid to help in resisting international communist aggression. The presentation used the phrase "international communism," which made the doctrine much broader than simply responding to Soviet military action. A danger that could be linked to communists of any nation could conceivably invoke the doctrine. The doctrine was absolute and did not establish any side rails or processes to implement that support. US troops in Vietnam during 1961 were 3,201.

The Kennedy Doctrine

Kennedy's Doctrine refers to the foreign policy initiatives of John Fitzgerald Kennedy. Generally, Kennedy expanded the policies of containment of communism outlined by his predecessors, President Eisenhower and President Truman. In his Inaugural Address, President Kennedy made a special pledge to our southern neighbors by stating:

> To our sister republics south of our border, we offer a special pledge—to convert our good words into good deeds—in a new alliance for progress—to assist free men and free governments in casting off the chains of poverty. But this peaceful revolution of hope cannot become the prey of hostile powers. Let all our neighbors know that we shall join

with them to oppose aggression or subversion anywhere in the Americas. And let every other power know that this Hemisphere intends to remain the master of its own house.

He effectively enlarged the Monroe Doctrine to include our entire Hemisphere. He also called upon the public to assist in "a struggle against the common enemies of man: tyranny, poverty, disease, and war itself."

Kennedy voiced support for the containment of communism and the reversal of communist progress in the Western Hemisphere. In his 1952 address to the Massachusetts Chapter of the American Federation of Labor, Kennedy stated that communists are "an enemy, powerful, unrelenting and implacable who seeks to dominate the world by subversion and conspiracy." He asserted that "all problems are dwarfed by the necessity of the West to maintain against the communists a balance of power." US troops in Vietnam during 1963 numbered 16,300.

The Johnson Doctrine

President Johnson continued his post–World War II predecessor's commitment to the containment of communist expansion. It has been reported that Johnson specifically stated that domestic revolution in the Western Hemisphere would no longer be a local matter when the object is the establishment of a "Communist dictatorship". This statement builds off the fundamentals of the Monroe Doctrine. But Johnson went further when he stated his opposition to permitting democratic East Asian Nations to fall to communist takeovers.

US troops in Vietnam during 1969 numbered 475,200.

The Nixon Doctrine

On July 25, 1969, at a press conference in Guam, Richard Nixon introduced the Nixon Doctrine. Nixon stated, "The United States will participate in the defense and development of allies and friends, but that America cannot—and will not—conceive all the plans, design all the programs, execute all the decisions, and undertake all the defense of the free nations of the world."

This doctrine reflects his belief that nations in Asia should have more responsibility for their own defense and security rather than relying solely on the United States.

He further added. "We, of course, will keep the treaty commitments that we have, but as far as our role is concerned, we must avoid the kind of policy that will make countries in Asia so dependent upon us that we are dragged into conflicts such as the one we have in Vietnam."

Nixon changed the fundamental criteria on how we would support our allies. On November 3, 1969, Nixon announced his doctrine, which he called "Peace through a Partnership." He outlined the Doctrine with the following three points:

- First, the United States will keep all its treaty commitments.
- Second, we shall provide a shield if a nuclear power threatens the freedom of a nation allied with us or of a nation whose survival we consider vital to our security.
- Third, in cases involving other types of aggression, we shall furnish military and economic assistance when requested in accordance with our treaty commitments. But we shall look to the nation directly threatened to assume the primary responsibility of providing the manpower for its defense.

The doctrine was also applied by the Nixon administration in the Persian Gulf region, with military aid to Iran and Saudi Arabia, so that these US allies could undertake the responsibility of ensuring peace and stability in the region. The United States first had a weak treaty with Saudi Arabia in 1931. The first treaty with Iran was entered into on August 15, 1955. The Nixon Doctrine, which created the Vietnamization of the war in Vietnam—a program designed to shift the responsibility of the war from the United States to the South Vietnamese, allowing the United States to gradually withdraw its troops from Vietnam—reduced the number of US soldiers in Vietnam from over 500,000 in 1969 when Nixon took office to 50,000 in 1973 when Ford took office.

People wonder if this was the start of countries beginning to believe that the United States was not a trusted ally and the foundation of the US Gulf difficulties and eventuality of the Gulf wars.

The First 25 Years After World War II

It is clear after reviewing all the military activities from Greece until the start of the First Gulf War that no administration looked at any criteria other than curtailment of the growth of communist activities.

If we had simply followed the Powell Doctrine, I believe our human and financial costs would have been drastically reduced.

The Carter Doctrine

Proclaimed by President Jimmy Carter in his State of the Union Address on January 23, 1980, Carter stated that the United States would use military force if necessary to defend its national interests in the Persian Gulf region. The doctrine was a response to the 1979 invasion of Afghanistan by the Soviet Union and was intended to deter the Soviet Union— the Cold War adversary of the United States—from seeking hegemony in the Persian Gulf. After stating that Soviet troops in Afghanistan posed "a grave threat to the free movement of Middle East oil," Carter proclaimed: "Let our position be clear. An attempt by any outside force to gain control of the Persian Gulf region will be regarded as an assault on the vital interests of the United States of America, and such an assault will be repelled by any means necessary, including military force."

This, the key sentence of the Carter Doctrine, was written by Zbigniew Brzezinski, President Carter's National Security Adviser. Brzezinski modeled the wording of the Carter Doctrine on the Truman Doctrine and insisted that the sentence be included in the speech "to make it very clear that the Soviets should stay away from the Persian Gulf."

In *The Prize: The Epic Quest for Oil, Money, and Power*, author Daniel Yergin notes that the Carter Doctrine "bore striking similarities" to a 1903 British declaration, in which British Foreign Secretary Lord Landsdowne warned Russia and Germany that the British would "regard the establishment of a naval base or of a fortified port in the Persian Gulf by any other power as a very grave menace to British interests, and we should certainly resist it with all the means at our disposal."

The Reagan Doctrine

An important Cold War strategy by the United States was to oppose the influence of the Soviet Union by backing anticommunist guerrillas against the communist governments of Soviet-backed client states. This strategy, partially created in response to the Brezhnev Doctrine, was a centerpiece of American foreign policy from the mid-1980s until the end of the Cold War in 1991.

Reagan first explained the doctrine in his 1985 State of the Union Address: "We must not break faith with those who are risking their lives on every continent, from Afghanistan to Nicaragua—to defy Soviet aggression and secure rights which have been ours from birth. Support for freedom fighters is self-defense."

The Reagan Doctrine called for American support of the Contras in Nicaragua, the Mujahideen in Afghanistan, and Jonas Savimbi's UNITA movement in Angola, among other anticommunist groups.

Bush 41 and Bush 43 Doctrines

The "Bush Doctrine" typically refers to the foreign policy principles associated with President George W. Bush, who served from 2001 to 2009. However, the first President Bush, George H.W. Bush (often called "Bush 41" to distinguish him from his son, "Bush 43"), also had his own set of foreign policy principles during his presidency from 1989 to 1993. His doctrine, often called the "Bush 41 Doctrine," emphasized a few key elements:

Key Elements of the Bush 41 Doctrine

- **Multilateralism:** George H.W. Bush placed a strong emphasis on working with international allies and through organizations like the United Nations. He believed in building coalitions and promoting international cooperation, as seen during the Gulf War (1990–1991). He assembled a broad coalition to oppose Iraq's invasion of Kuwait, gaining support from NATO allies, Middle Eastern nations, and the United Nations.

- **New World Order:** Bush envisioned a "new world order" after the end of the Cold War. This concept involved a world where the United States and its allies would work together to maintain peace and stability based on international law and collective security. The idea aimed to transition away from the Cold War's bipolar power struggle and toward a more cooperative global system.
- **Peaceful Resolution of Conflicts:** Bush 41 sought to resolve conflicts diplomatically where possible. For example, he pursued arms reduction treaties with the Soviet Union (later Russia), such as START I and the Chemical Weapons Convention.
- **Humanitarian Interventions:** Though cautious, Bush 41 occasionally supported humanitarian interventions, as seen in Somalia (1992), where the United States intervened to provide relief and restore order amid a humanitarian crisis.
- **Containment and Stability:** In line with Cold War doctrines, Bush continued policies of containment where necessary, but he aimed for stability rather than confrontation. His administration supported the reunification of Germany within NATO, signaling a preference for peaceful transitions over aggressive stances.

The Bush 41 Doctrine is often remembered for its emphasis on multilateralism, diplomacy, and coalition-building, contrasting with the more unilateral and preemptive approach associated with the Bush 43 Doctrine. Also, Bush 41 was the president when the Powell Doctrine was used so successfully.

The Clinton Doctrine

Unlike other presidential doctrines, Clinton's was not a clear statement. However, in a February 26, 1999, speech, President Bill Clinton said the following, which was considered the Clinton Doctrine:

> It's easy to say that we really have no interest in who lives in this or that valley in Bosnia or who owns a strip of brushland in the Horn of Africa or some piece of parched earth by the Jordan River. But the true measure of our interests lies not in how small or distant these places are or in whether we have trouble pronouncing their names. The question we must ask is, what are the consequences to our security of letting conflicts fester and spread? We cannot—indeed, we should not—do everything or be everywhere. But where our values and our interests are at stake and where we can make a difference, we must be prepared to do so.

Later statements, "genocide is in and of itself a national interest where we should act," and "we can say to the people of the world, whether you live in Africa, or Central Europe, or any other place, if somebody comes after innocent civilians and tries to kill them en masse because of their race, their ethnic background or their religion, and it's within our power to stop it, we will stop it," augmented the doctrine of interventionism. However, a lack of leadership within the Clinton administration led to the United States' failure to intervene effectively in the Rwandan genocide despite this ideology.

The Bush Doctrine (Bush 43)

The Bush Doctrine is the set of foreign policies adopted by the president of the United States, George W. Bush, in the wake of the September 11, 2001, attacks. In an address to the United States Congress after the attacks, President Bush declared that the United States would "make no distinction between the terrorists who committed these acts and those who harbor them," a statement that was followed by the US invasion of Afghanistan. Subsequently, the Bush Doctrine has come to be identified as a policy that permits preventive war against potential aggressors before they are capable of mounting attacks against the United States, a view that has been used in part as a rationale for the 2003 Iraq War.

The Bush Doctrine is a marked departure from the policies of deterrence that generally characterized American foreign policy during the Cold War and the brief period between the collapse of the Soviet Union and 9/11. It can also be contrasted with the Kirkpatrick Doctrine of supporting stable right-wing dictatorships, which was influential during Ronald Reagan's administration.

Here are the Main Tenets of the Bush 43 Doctrine

1. **Preemptive Action:**
   The Bush Doctrine emphasized the right of the United States to take **preemptive action** against perceived threats, particularly terrorist groups and rogue states that might acquire weapons of mass destruction (WMDs). This represented a significant shift from the traditional policy of deterrence. The doctrine argued that waiting for an imminent threat to materialize fully was too risky, especially in an era of terrorism and WMDs. Thus, the United States reserved the right to strike first to neutralize potential threats.

2. **Unilateralism:**
    The Bush administration advocated for **unilateral action** when necessary, stating that the United States would act alone if allies were unwilling or unable to participate in efforts to combat terrorism and other security threats. This approach reflected a willingness to bypass international institutions like the United Nations if they were seen as ineffective or slow to respond, as demonstrated in the lead-up to the Iraq War.

3. **Spreading Democracy:**
    The Bush Doctrine promoted **democracy promotion** as a cornerstone of US foreign policy, particularly in the Middle East. The administration argued that spreading democracy would help combat terrorism by addressing the root causes of extremism, such as authoritarianism and lack of political freedom. The belief was that democratic nations would be more peaceful and less likely to harbor terrorists, which led to efforts to establish democratic governments in Afghanistan and Iraq. However, Bush 43, along with most other presidents, ignored historical governmental systems, some of which had been in place for hundreds of years, in favor of forcing a change to a Western style of government.

4. **The War on Terror:**
    Central to the Bush Doctrine was the **Global War on Terror**, which involved a broad and open-ended campaign against terrorist groups, particularly al-Qaeda, and any state that supported or harbored terrorists. This led to military interventions in Afghanistan (2001) to dismantle al-Qaeda and remove the Taliban from power, and later in Iraq (2003)

based on claims that Saddam Hussein possessed WMDs and had ties to terrorist groups.

5.   **"You're Either with Us or Against Us":**
     The Bush Doctrine included a clear message to other nations: they were either aligned with the United States in its fight against terrorism or considered hostile if they opposed or failed to cooperate. This stance was meant to pressure countries to actively support counterterrorism efforts and cut off support for terrorist organizations.

Impact and Criticism

The Bush Doctrine marked a departure from previous US foreign policies, embracing a more aggressive and interventionist stance. It faced criticism for:

- Leading to prolonged conflicts in Afghanistan and Iraq with significant costs and loss of life.
- Damaging international alliances and the reputation of the United States due to perceived unilateralism.
- Creating unintended consequences, such as regional instability and the rise of extremist groups like ISIS.

As I mentioned, the Bush Doctrine prioritized preemptive action, unilateralism, democracy promotion, and an uncompromising stance on the War on Terror. It defined US foreign policy during Bush 43's presidency and had lasting effects on global geopolitics. All of these actions conflicted with the Powell Doctrine.

The Obama Doctrine

The Obama Doctrine is yet to be fully defined, and President Obama himself has expressed a dislike for an overly "doctrinaire" approach to foreign policy. When asked about his doctrine, Obama has replied that the United States would have to "view our security in terms of a common security and a common prosperity with other peoples and other countries." On April 16, 2009, E. J. Dionne wrote a column for *The Washington Post* defining the doctrine as "a form of realism unafraid to deploy American power but mindful that its use must be tempered by practical limits and a dose of self-awareness." The Obama Doctrine has been praised by some as a welcome change from the dogmatic and aggressive Bush Doctrine.

Others, such as Bush appointee and former US Ambassador to the United Nations John Bolton, have criticized it as overly idealistic and naïve, promoting appeasement with the country's enemies.

The Trump Doctrine

The Trump Doctrine is defined as the Trump administration's foreign policy, based upon the slogan of "America first." It leverages America's economic and military power to increase and decrease tensions favorably for America.

Trump was especially critical of so-called "free riders," or countries that the United States uses resources to protect without receiving benefits in return. Through his foreign policy, Trump criticized the use of US military forces in situations where national interests were uninvolved.

The Biden Doctrine

Although the Biden Doctrine is not explicitly defined, President Biden's foreign policy has been characterized by an avoidance of aggressive tactics that involve personnel in foreign nations.

As a means of moving away from the Trump Doctrine's policy of "America first," Biden stated that, "The transatlantic alliance is back. The US is determined to consult with you." The Biden Doctrine is a shift away from foreign conflicts to focus resources on domestic issues. Following the United States' withdrawal of troops from Afghanistan, Biden defended the decision as furthering the administration's efforts to reduce risk to American lives and resources in foreign affairs.

Doctrines of Secretaries of Defense

The McNamara Doctrine
Robert Strange McNamara was an American business executive and public official, best known for serving as the US Secretary of Defense from 1961 to 1968 under presidents John F. Kennedy and Lyndon B. Johnson. He played a significant role during the Cold War era, particularly in the Vietnam War and in shaping US defense policy.

McNamara's approach to defense policy, particularly in the context of nuclear strategy, became known as the "McNamara Doctrine." Key elements of this doctrine included the following:

Flexible Response: In contrast to the previously dominant strategy of "Massive Retaliation," which relied on an all-or-nothing nuclear response, McNamara advocated for a range of military options. He believed the United States should have multiple choices in its response to threats, from conventional forces to limited nuclear strikes, to avoid unnecessary escalation.

Mutual Assured Destruction (MAD): McNamara emphasized the concept of Mutual Assured Destruction as a deterrent against nuclear war. He argued that the United States and the Soviet Union must maintain credible second-strike capabilities so that any nuclear attack by one side would ensure the complete destruction of both. This theory was aimed at deterring either side from launching a first strike.

Quantitative Analysis and Systems Analysis: McNamara applied a data-driven, analytical approach to defense planning. He used statistical analysis and systems theory to make decisions about defense spending and military capabilities, aiming to maximize efficiency and effectiveness.

Gradual Escalation: During the Vietnam War, McNamara believed in a strategy of gradual escalation, where the United States would slowly increase its military presence and pressure on North Vietnam rather than launching a full-scale invasion. This approach was intended to demonstrate resolve while avoiding over-commitment, but it ultimately led to a prolonged and costly conflict.

While his policies shaped US defense strategy during a critical period, McNamara's legacy remains controversial, particularly due to his role in the Vietnam War. Later in life, he expressed regret over his involvement in the war, acknowledging the failures of his approach. His reflections are documented in the book *In Retrospect: The Tragedy and Lessons of Vietnam* and the documentary *The Fog of War*. The McNamara and Rumsfeld Doctrines were much alike in that they proclaimed technology and minimal forces would win the day.

The Rumsfeld Doctrine

Named after former US Secretary of Defense Donald Rumsfeld, the Rumsfeld Doctrine was a military strategy that emphasized using a smaller, highly agile force equipped with advanced technology to achieve quick, decisive victories. This doctrine stood in contrast to traditional military strategies that relied on large numbers of troops and heavy equipment.

Key Principles of the Rumsfeld Doctrine Include

1. **Light and Agile Forces:** The doctrine advocates for a smaller number of ground troops who can move quickly and adapt to changing battlefield conditions.
2. **High-Technology and Precision Weapons:** It places a strong emphasis on utilizing cutting-edge technology, including precision-guided munitions, drones, and real-time intelligence systems, to enhance the effectiveness of a smaller force.
3. **Speed and Surprise:** By deploying swiftly and striking with precision, the doctrine aims to achieve shock and awe, overwhelming the enemy before they can mount an effective defense.
4. **Reliance on Special Operations:** The doctrine incorporates a significant role for special operations forces who can conduct targeted missions and gather critical intelligence.

The Rumsfeld Doctrine was most notably applied during the early stages of the Afghanistan and Iraq wars. While it achieved initial successes in terms of rapid victories, the doctrine faced criticism for not adequately planning for long-term occupation and stability, particularly in Iraq. Critics argued that the reduced troop numbers left the United States unable to effectively manage insurgencies and rebuilding efforts after the major combat operations had concluded.

Powell and Shinseki argued that Rumsfeld's Doctrine would win the battle but ultimately lose because there would be insufficient troops to keep the peace. Powell was correct.

Bush 41—The Powell Doctrine Applied

George Bush, the 41st president of the United States, focused primarily on foreign policy during his time in the White House. During his presidency, Germany was reunifying, the Soviet Union was collapsing, and the Cold War was coming to an end. He improved US–Soviet relations, notably through a meeting with Soviet leader Mikhail Gorbachev, where they signed the Strategic Arms Reduction Treaty in July 1991.

Bush also authorized military operations in Panama and the Persian Gulf. In December 1989, the United States invaded Panama to overthrow dictator Manuel Noriega as president. Noriega was in control of the Panama Canal and was simultaneously running a major drug tariffing operation. Panama had a large population of expat Americans, and he presented a significant threat to American citizens living in Panama.

In October 1989, Bush appointed General Powell as chairman of the Joint Chiefs of Staff. At 52, he was one of the youngest chairmen in history. General Powell is regarded as one of the most effective senior officers in American history.

In August 1990, 10 months after General Powell's appointment as chairman, Iraqi leader Saddam Hussein invaded Kuwait and threatened Saudi Arabia. This effectively threatened the free flow of a significant portion of the world's oil supply.

During the buildup to the US response to the Invasion, the United States utilized the Powell Doctrine to prepare for the war. As already explored, Powell's doctrine has eight significant components that must be answered in the affirmative before we deploy the US military in a war. Again:

1. Is a vital national security interest threatened?
2. Do we have a clear, attainable objective?
3. Have the risks and costs been fully and frankly analyzed?
4. Have all other nonviolent policy means been fully exhausted?
5. Is there a plausible exit strategy to avoid endless entanglement?
6. Have the consequences of our action been fully considered?
7. Is the action supported by the American people?
8. Do we have genuine broad international support?

Powell has also asserted that "when a nation is engaging in war, every resource and tool should be used to achieve decisive force against the enemy, minimizing casualties and ending the conflict quickly by forcing the weaker force to capitulate."

Utilizing the Powell Doctrine, President Bush organized a coalition of over 30 countries. This coalition launched a US-led air assault against Iraq in January 1991. After five weeks of air strikes and less than 100 hours of ground combat, Operation Desert Storm concluded with Iraq's defeat and Kuwait's liberation. General Powell's Doctrine helped ensure the coalition suffered minimal casualties while defeating one of the world's largest armies. Some experts say that this is the only military operation since World War II that the United States and its allies have won.

Drawing a side-by-side comparison between the Powell and McNamara Doctrines, we observe the following:

Decisiveness vs. Gradualism: The Powell Doctrine advocates for quick, overwhelming force, while the McNamara Doctrine favors gradual escalation.

Public Support: Powell emphasized the need for broad public support, whereas McNamara's approach sometimes led to actions that were unpopular and eroded public trust.

Clear Objectives and Exit Strategy: The Powell Doctrine insists on defined objectives and an exit strategy, while the McNamara Doctrine's flexible response and focus on containment often led to open-ended engagements.

Drawing a side-by-side comparison between Powell and Rumsfeld, I believe that you will observe that Rumsfeld and McNamara's approaches were similar. We observe the following:

Decisiveness vs. Gradualism: The Powell Doctrine advocated for quick, overwhelming force, while the Rumsfeld Doctrine, against the Chief of Staff's recommendation of using 500,000 troops, started with 150,000 troops and gradually escalated the forces to 393,000.

Public Support: Powell emphasized the need for broad public support, whereas Rumsfeld's approach started with support, but sometimes actions led to unpopular outcomes and eroded public trust.

Clear Objectives and Exit Strategy: The Powell Doctrine insists on defined objectives and an exit strategy, while the Rumsfeld Doctrine's ignores this objective. It focuses on flexibility, technology, and speed.

Minimizing Casualties: One must assume that Rumsfeld would want to minimize casualties, whereas the Powell Doctrine specifically states that is an objective. The following table shows the number of soldiers killed and wounded by enemy action in Gulf 1 (Powell Doctrine) and Gulf 2 (Rumsfeld Doctrine):

	Gulf 1	Gulf 2
Killed in action	145	4,500
Wounded in action	467	32,000

Ending the Conflict Quickly: Following the Powell Doctrine, a principal objective of the First Gulf War was to end the conflict quickly by forcing the weaker force to capitulate. On the day ground operations started in the First Gulf War, the collation had approximately 750,000 total forces, including 540,000 US forces. The Second Gulf War planning process started with a public disagreement between the US Army's Chief of Staff, General Eric K. Shinseki, and Secretary Rumsfeld. Secretary Rumsfeld said it would take approximately 150,000 troops to win the war while General Shinseki's position was that it would take around 500,000 troops to win the war and the peace. As a result of this disagreement, General Shinseki retired before the commencement of the attack.

The war commenced with approximately a couple hundred thousand troops, but because of some setbacks caused by an insurgency, US troops peaked at around 340,000 US troops and 45,000 British troops.

The following table reflects what was reported by publications (I believe that the chart understates the length of time for the Second Gulf War):

	Gulf 1	Gulf 2
Ground Combat	100 Hours	2,752 Days

THE RUMSFELD MISTAKES

The "Rumsfeld Doctrine" is a phrase coined by journalists concerned with the perceived transformation of the US military. Named after former US Secretary of Defense Donald Rumsfeld, the Doctrine would be considered Rumsfeld's own take on RMA (Revolution in Military Affairs). It seeks to increase force readiness and decrease the amount of supply required to maintain forces by reducing the number in a theater. This is done mainly by using light-armored vehicles (LAVs) to scout for enemies who are then destroyed via airstrikes. The basic tenets of this military strategy are

- High-technology combat systems
- Reliance on air forces
- Small, nimble ground forces

The early phases of the Afghanistan and Iraq wars are considered the two closest implementations of this doctrine.

Secretary Rumsfeld had a contentious relationship with several members of America's senior military leadership. General

Shinseki (US Army Chief of Staff) wanted to follow the Powell Doctrine and have approximately 500,000 troops available to defeat Iraq and maintain the peace. Secretary Rumsfeld believed we only needed 150,000 troops to "win" the war. General Shinsaku believed that while 150,000 troops might prevail in the initial battle, that number would be wholly insufficient to win the peace.

In retrospect, it is very clear that General Shinseki was correct. Shortly after Iraq was conquered, a 19-year insurgency started. The insurgency occurred for two principal reasons:

- First, the Coalition did not have sufficient resources to maintain a peaceful environment.
- Second, the Coalition Provisional Authority (CPA) issued CPA Order 1, which banned all senior members of the Ba'ath Party from the new government, public schools, and colleges.

i. The concept was known as De-Ba'athification of Iraqi Society. The process was disclosed in May 2003 when Paul Bremer issued CPA Orders to exclude from the new Iraq the top four levels of the Baath Party (CPA Order 1). For instance, this decision effectively eliminated the leadership and top technical capacity for universities, hospitals, transportation, electricity, and communications.[6] As an example, in the Health Ministry, a third of the staff were forced out, and eight of the top twelve officers in the organization were excluded. This also included "forty thousand schoolteachers, who had joined the Baath Party simply to keep their jobs."[7]

ii. On May 23, 2003, CPA Order Number 2 was issued. This order disbanded the Iraqi military, security, and intelligence

---

[6]Chandrasekaran, Imperial Life, p.82.

[7]George Tenet, At the Center of the Storm: My Years at the CIA (New York: HarperCollins 2007) p.427.

infrastructure (CPA Order 2). Disbanding the armed forces and the police resulted in 670,000 individuals immediately losing their jobs. This act created a large pool of armed and trained individuals who were almost immediately hostile to the coalition forces.

iii. These policies were contrary to the policies of the US Military in both the European Theater of Operations and the Pacific Theater of Operations after World War II. While at the end of World War II the European Theater of Operations initially had a similar policy, they quickly dropped it when they saw the damage to the reconstruction and safety concerns it was going to cause.

Both of these theaters' final policies—which had no universal ban on employment—resulted in:

1. The European Theater reduced the risk of political extremism and laid the groundwork for long-term economic recovery and political stability. By 1958, Germany's industrial production was four times higher than it was a decade earlier.[8]

2. Japan transitioning to a stable democracy that experienced rapid economic growth, becoming a major global economy by the 1960s.

Political Strategy: Encouraged the establishment of a democratic government but faced significant challenges due to sectarian divisions and insurgency. The CPA rules were extremely complex to implement and laid the ground for a decade-plus of insurgency. This occurred because of the very large number of unemployed individuals as a result of the implementation of

---

[8]Foundation for Economic Education.org. "The German Economic Miracle and the 'Social Market Economy."

CPA. In addition, CPA1 and CPA2 were contrary to the successful policies of the US Military in both the European Theater of Operations and the Pacific Theater of Operations after World War II.

Outcome: The approach faced criticism for its execution, leading to prolonged instability, violence, and a difficult transition to a stable government. Iraq experienced a rise in sectarian violence and the emergence of extremist groups, undermining efforts to establish democracy.

Many individuals joined the Ba'ath Party because it was a prerequisite to be a member of the Party to have a good job. Banning all the former members of the Ba'ath Party, including those who had no meaningful relationship with the party, was a grave mistake. The CPA orders fueled the flame to magnify the insurgency because all former military officers and enlisted individuals, policemen, and anyone who was employed to run the country's infrastructure were prohibited from employment with the new government. This built resentment and provided many trained individuals to operate the insurgency. The Ba'ath Party's insurgency used the same methods that the VC used against America's soldiers in the Vietnam War.

Their forces avoided head-on military confrontation and resorted to political subversion and guerrilla tactics.

This mistake was very similar to some initial mistakes that were made in Germany after World War II. In JCS Directive 1067, General Eisenhower received the following order:

All members of the Nazi party who have been more than nominal participants in its activities, all active supporters of Nazism or militarism, and all other persons hostile to Allied purposes will be removed and excluded from public office and from positions of importance in quasi-public and private enterprises. No such persons shall be retained in any of the categories of employment listed above because of administrative necessity, convenience, or expediency.

However, the Allied powers sometimes disregarded the denazification order for rocket scientists, technical engineers, and other specifically skilled individuals. To some, it seemed that anyone who could provide technical assistance to the US military was automatically not considered a Nazi. At the end of 1945, over 3 million Germans were on a waiting list to determine if they could be de-nazified. The list was growing by as much as 40,000 applications a day. In early 1946, a law was passed that provided civilian tribunals with the ability to review cases. This move was approved and expedited by the German people. By 1947, the number of Nazis held in detention had dropped to 90,000.

Those mistakes, however, were not made in Japan, where General Douglas McArthur implemented a program known as the "Reverse Course." Most individuals who wanted to re-enter national programs were reviewed for their quality. Over time, thousands of nationalist wartime leaders were de-purged and allowed to re-enter politics and government ministries. In retrospect, this is why banning all the former members of the Ba'ath Party, including those who had no meaningful relationship with the party, was a grave mistake.

This mistake was very similar to the initial mistakes that were made in Germany after World War II. If Rumsfeld had not banned the white hat members of the Ba'ath Party, most individuals believe that the insurgency that caused the deaths of so many Americans would not have happened.

THE MARSHALL DOCTRINE

President Franklin Delano Roosevelt remarked to General Dwight Eisenhower one day during World War II, "I hate to think that fifty years from now practically nobody will know who George Marshall was."

Who was General George Marshall? He became chief of the Army in 1939, ruthlessly purged the ranks of his generals, and set an enduring standard for what it takes to lead U.S. troops. "The present general officers of the line are, for the most part, too old to command troops in battle under the terrific pressure of modern war," Marshall said in October 1939, a month after being sworn in as chief. "Most of them have their minds set in outmoded patterns and can't change to meet the new conditions they may face if we become involved in the war that's started in Europe."

One of the major prewar military exercises was the Bib Louisiana Maneuvers. The Maneuvers were staged during August and September 1941. Marshall was demanding and held his subordinates to very high standards. For instance, he decided that the American officer corps needed quick thinkers who were resourceful and relentlessly determined. Officers also required an abundance of common sense, which would prevent the gross errors that stem from hasty decisions and actions. Only 11 of the 42 generals who commanded a division, a corps, or an army in the maneuvers would go on to command in combat.

Marshall removed officers in part to convey a sense of urgency. When the commandant at Leavenworth, Brigadier General Charles Bundel, told him that updating army training manuals would take 18 months, Marshall offered him three months. No, it can't be done, Bundel responded. Marshall then offered four months. Bundel again said it was impossible. Marshall asked him to reconsider. "You be very careful about that," Marshall warned.

"No, it can't be done," Bundel insisted.

"I'm sorry, then you are relieved," Marshall said.

Thomas E. Ricks wrote a great article on the subject called **Failure is Not an Option: George C. Marshall's Relentless Pursuit of Leadership Excellence**. To paraphrase the much longer article:

During World War II, General George C. Marshall revolutionized the U.S. military by enforcing a ruthless standard of leadership. From the outset, he purged ineffective officers, including dozens of generals deemed unfit for the modern battlefield. This approach was influenced by his World War I experiences under General John Pershing, who also demanded results and swiftly removed underperformers.

Marshall's philosophy emphasized character over intellect, seeking leaders who were decisive, energetic, and adaptable under pressure. He rejected "calamity howlers"—pessimists who sapped morale—and avoided promoting reckless or flamboyant individuals. Instead, he prioritized team players with sound judgment and a sense of duty.

Under Marshall's guidance, the army grew from 197,000 personnel to over 8 million by 1945. This transformation involved identifying and promoting capable leaders while retiring those unable to meet wartime demands. His meritocratic system rewarded initiative and punished failure, creating an environment where prudent risk-taking was encouraged. Many leaders, such as Dwight Eisenhower, rose rapidly under his system, proving its effectiveness.

Though sometimes harsh, Marshall's leadership doctrine reshaped the U.S. Army into a modern force. His insistence on rigorous standards and his willingness to challenge entrenched norms ensured the military's readiness and success during the war.

While occasionally perceived as brutal, the Marshall system generally produced military effectiveness. Dwight Eisenhower offers the proof. Just a year before the start of World War II, he was still a lieutenant colonel, not even in command of a regiment. Yet because Marshall saw in him the qualities of a good leader, Eisenhower, within a few years, was commanding armies of millions.

Decades later, Eisenhower recalled how Marshall moved against so many top officers following the Louisiana Maneuvers. "By God, he just took them and threw them out of the room, but ultimately," Ike concluded, "Marshall was vindicated. He got them out of the way, and I think, as a whole, he was right to do so."

THE THREE PLANS TO DEAL WITH COUNTRIES
THE U.S. CONQUERED

This section compares The Marshall Plan for governance in Europe following World War II, MacArthur's governance plan for post-World War II operations in Japan, and Donald Rumsfeld's approach to post-combat operations in Iraq. Two of the approaches were highly successful and represent different strategies for rebuilding and stabilizing nations after conflict. Here's a comparative analysis of these three approaches:

Marshall Plan (1948)

Overview:
Context: Germany was occupied but was partitioned by the four Allied forces after its defeat in World War II. Implemented after World War II to aid Western European countries in recovery. I believe that was the objective of the US, GB, and maybe France. The Russians stripped Germany of industrial

machinery, priceless, art, and almost anything else of value as reparations for the damages they incurred because of the war.

Goals: To prevent the spread of communism, rebuild war-torn economies, and restore political stability.

Approach:

1. Financial Aid: Provided over $13 billion (equivalent to over $100 billion today) in economic assistance to help rebuild European economies.

2. Economic Integration: Encouraged trade and cooperation among European nations to foster economic interdependence.

3. Political Stability: Aimed to bolster democratic governments and counter communist influence.

Treatment of the Ruling Class Under Marshall's Plan

Collaboration with Existing Structures:

The Marshall Plan aimed to stabilize and rebuild war-torn economies, which necessitated collaboration with existing political and economic elites. The United States recognized that established leaders were often essential for effective governance and reconstruction.

The plan provided aid to governments that were committed to democratic reforms, thus promoting stability through the involvement of moderate and pragmatic elites.

Support for Democratic Governance:

The United States encouraged democratic political systems in Europe as a bulwark against the spread of communism. This meant working with ruling classes that were willing to embrace democratic principles and reforms.

Leaders who supported the Marshall Plan were often those who had been part of pre-war political establishments but had adapted to the postwar context, promoting stability and recovery.

Economic Incentives:

The financial aid provided by the Marshall Plan was conditional on the implementation of economic and political reforms. This created an environment where ruling classes had to adapt to receive aid, promoting a shift toward more cooperative and democratic governance.

Countries had to demonstrate sound economic policies and a commitment to democratic principles to qualify for assistance, which encouraged ruling elites to align with these goals.

Encouragement of Integration:

The Marshall Plan promoted economic cooperation among European nations, which required ruling classes to work together across national borders. This helped create a sense of shared interests among elites, fostering a collaborative approach to recovery.

The initiative served as a precursor to efforts that would ultimately lead to greater European integration, influencing the behavior of ruling classes toward collaboration rather than competition.

Focus on Stability Over Radical Change:

The Marshall Plan did not seek to dismantle existing ruling classes outright but rather to stabilize them through economic recovery and political reform. This pragmatic approach aimed to prevent the rise of extremist movements by supporting moderate and democratic leaders.

By bolstering existing political structures, the plan helped maintain order and stability during a period of potential upheaval.

Long-Term Impact on Governance:

The success of the Marshall Plan contributed to the establishment of strong democratic institutions in Western Europe, which in turn solidified the position of moderate political elites.

Over time, the ruling classes that adapted to the new democratic environment were able to maintain their influence, though their power was increasingly subject to democratic accountability.

Outcome:

Successfully revitalized European economies, leading to sustained growth and political stability, and is often credited with laying the groundwork for the European Union.

The EU came about after Russia failed in 1991. The European Union—a geopolitical entity created in 1993, covering much of the European continent—was founded upon numerous treaties and has undergone expansions and secessions that have taken it from six member states to 27, a majority of the states in Europe. The institutionalization and integration of modern Europe began in 1948.

MacArthur's Governance of Japan (1945–1951)

Overview:

Context: Japan was occupied by Allied forces after its defeat in World War II.

Goals: To demilitarize and democratize Japan, promote economic recovery, and ensure a stable, peaceful society.

Approach:

Political Reforms: Implemented a new constitution emphasizing democracy, civil rights, and women's suffrage.

Economic Reforms: Land reforms redistributed land from landlords to tenants, promoting agricultural productivity. MacArthur's administration implemented land reforms that redistributed land from large landowners to tenant farmers, undermining the traditional agricultural elite.

Education and Cultural Changes: Reformed the education system and promoted democratic value

Treatment of Ruling Class Under Macarthur's Plan:

1. Inclusion of some elites: While he purged many from power, MacArthur also recognized the need for stability and sometimes retained certain bureaucrats who were deemed necessary for effective governance, particularly in the education and business sectors.

2. Promotion of Democratic Leaders: He encouraged the emergence of new political leaders who supported democratic governance, fostering a new political elite aligned with US interests.

3. Demilitarization and Democratization: MacArthur aimed to dismantle Japan's militaristic ruling class. He purged former military leaders and government officials associated with the wartime regime.

4. Land Reforms: MacArthur's administration implemented land reforms that redistributed land from large landowners to tenant farmers, undermining the traditional agricultural elite.

Outcome:

Japan transitioned to a stable democracy and experienced rapid economic growth, becoming a major global economy by the 1960s.

Rumsfeld's Plan for Post-Combat Iraq (2003–2006)

Overview:

Context: Following the invasion of Iraq in 2003, after the overthrow of Saddam Hussein.

Goals: "He didn't want there to be any stability operations—also known as "nation-building." And so, the United States went into the war without any such plan. This wasn't an oversight; it was deliberate—it was Rumsfeld's plan to have no plan. Catastrophe ensued: Rumsfeld seemed to believe that you could increase force readiness and decrease the amount of supply required to maintain forces, by reducing the number in a theater. This is done mainly by using light armored vehicles (LAVs) to scout for enemies who are then destroyed via airstrikes. The basic tenets of this military strategy are:

i.   High-technology combat systems
ii. Reliance on air forces
iii. Small, nimble ground forces

Approach:

Military Focus: Emphasized a rapid military victory followed by a limited post-combat role for US forces.

This approach resulted in insufficient resources to achieve the stated objective. The initial phase lacked adequate troop levels and resources for effective governance and reconstruction.

Another way of saying this is that his plan successfully conquered the country but incentivized the development of the insurrection.

Rumsfeld rejected the Chief of Staff's recommendation for 500,000 troops being necessary to maintain the peace post-combat operations, and as a result, Rumsfeld pared down the military's war plan for Iraq from 500,000 troops to just 140,000, which was clearly inadequate.

However, he was wrong in a way that he didn't think mattered but, in fact, mattered a great deal: those 140,000 troops were not nearly enough to hold the territory (the invading Americans would conquer a village, then move on to the next temporary objective, leaving chaos or worse in the dust) or to stave off the insurgency and civil war that followed and lasted for another nine years.

Treatment of Ruling Class Under Rumsfeld's Plan:

As described in the previous chapter in greater detail ("De-Baathification" of Iraqi Society), this decision effectively eliminated the leadership and top technical capacity for universities, hospitals, transportation, electricity, and communications. As an example, in the Health Ministry, a third of the staff were forced out, and eight of the top twelve officers in the organization were excluded. This also included "forty thousand schoolteachers, who had joined the Baath Party simply to keep their jobs."

Outcome:

The approach faced criticism for its execution, leading to prolonged instability, violence, and a difficult transition to a stable government. Iraq experienced a rise in sectarian violence and the emergence of extremist groups, undermining efforts to establish democracy.

Comparative Analysis

Goals:

All three initiatives aimed to promote **stability and democracy**, albeit in different contexts and with different immediate objectives.

Execution:

i. The **Marshall Plan** and **MacArthur's governance** employed comprehensive and well-resourced strategies that included economic, political, and social reforms.

ii. Rumsfeld's plan was criticized for its lack of a coherent, adequately resourced strategy post-invasion which led to instability and insurgency.

Outcomes:

The Marshall Plan and MacArthur's reforms are largely viewed as successful in achieving their goals, leading to stable democracies and economic recovery.

Rumsfeld's approach is often cited as a failure in terms of achieving long-term stability and security in Iraq, leading to ongoing conflict and unrest.

Furthermore, a quote from CNN on Thursday, August 15, 2024 stated: Iraq has postponed announcing an end date for the US-led military coalition's presence in the country due to recent developments, raising questions about the future of US military presence in the Gulf state amid heightened tension in the region.

Iraq's Higher Military Commission had aimed to propose an end date for Operation Inherent Resolve, the US military operation combatting the terror group ISIS.

"We were very close to announcing this agreement, but due to recent developments, the announcement of the end of the international coalition's military mission in Iraq was postponed," a statement by Iraq's foreign ministry said Thursday, without giving further details on what the recent developments included.

So, 20 years after the commencement of the Iraq operation, we will still have combat operations in Iraq because of regional tensions in the area.

Conclusion:

While the Marshall Plan and MacArthur's governance were characterized by comprehensive strategies that successfully fostered recovery and stability, Rumsfeld's post-combat Iraq plan illustrates the challenges of implementing democratic reforms in a complex and volatile environment without sufficient planning and resources.

THE UNITED STATES AND THE MODERN ISRAELI–ARAB WAR

Amerca's involvement in the modern Arab–Israeli conflict has been extensive and multifaceted, reflecting a mix of political, military, and economic interests. But let's explore the role our military and government have played in this controversial conflict:

1. **Support for Israel:** Since the establishment of Israel in 1948, the United States has been one of its strongest allies. This support includes extensive military aid, economic assistance, and diplomatic backing. The United States provides Israel with advanced military technology, intelligence cooperation, and substantial foreign aid, which has helped Israel maintain a technological edge in the region.

2. **Mediation Efforts:** The United States has often played the role of mediator in attempts to resolve the Arab–Israeli conflict. American presidents and diplomats have been involved in multiple peace processes, such as:

- **Camp David Accords (1978):** Brokered by President Jimmy Carter, this agreement between Egypt and Israel led to the first peace treaty between Israel and an Arab country.
- **Oslo Accords (1993):** Under President Bill Clinton's administration, these agreements were a significant step toward a two-state solution involving direct negotiations between Israel and the Palestinian Liberation Organization (PLO).
- **The Road Map for Peace (2003):** This plan, initiated by President George W. Bush, outlined steps for both Israelis and Palestinians to achieve a two-state solution.
- **Abraham Accords (2020):** These normalization agreements between Israel and several Arab countries, including the UAE and Bahrain, were facilitated by the Trump administration.

3. **Strategic Interests:** America's involvement is also shaped by its strategic interests in the Middle East, including:
   - **Energy Security:** Although the United States has become less dependent on Middle Eastern oil, stability in the region remains crucial for global energy markets and America's allies.
   - **Counterterrorism:** The United States has sought to limit the influence of groups like Hamas and Hezbollah, which oppose Israel and are supported by Iran. The United States also considers Iran's influence in the region a threat to both American and Israeli interests.
   - **Alliance with Arab States:** The United States maintains strong ties with key Arab states, such as Egypt, Saudi Arabia, and Jordan. Balancing these alliances with support for Israel has often been a delicate diplomatic task.

4. **Military Involvement:** While the United States has not directly fought in the Arab–Israeli wars, it has played an indirect role through arms sales, military training, and intelligence sharing. During the Cold War, the United States sought to counter Soviet influence in the region by supporting Israel and some Arab states.

5. **Aid to the Palestinians:** The United States has also provided humanitarian aid and development assistance to Palestinians, particularly through organizations like the United Nations Relief and Works Agency (UNRWA). However, this aid has often been a point of contention and has fluctuated depending on US policy and the political situation.

6. **Controversies and Criticism:** The United States has faced criticism for its policies, both domestically and internationally. Supporters of Israel appreciate America's backing, but critics argue that US support has sometimes been one-sided, undermining peace efforts and Palestinian rights. Additionally, decisions such as recognizing Jerusalem as Israel's capital in 2017 stirred significant controversy.

America's involvement in the Arab–Israeli conflict is a careful and controversial blend of supporting Israel, attempting to mediate peace, advancing strategic interests, and navigating regional alliances. This involvement has influenced the dynamics of the conflict and has had both stabilizing and destabilizing effects over the decades.

To give some context to how the Israeli–Arab wars fit into our timeline and help shape American–Israeli relations, I wanted to share a sequence of stories that Jerry Klinger articulates beautifully in the following passage.

Glenn King and Bill Gerson were killed on April 21, 1948, when their heavily overloaded C-46 Commando aircraft crashed on take-off at Mexico City airport while bound for Israel with a military cargo. Ironically, these two Americans were actually the IAF's first casualties.

"A Cobber" By Jerry Klinger

It took quite a while to find the gravesite where Glenn E. King's remains were resting. A few weeks back, I had never heard of him until I began reading details of the history of the nascent Israeli Air Force. On November 29, 1947, the United Nations created a Jewish state and an Arab state within the Palestinian mandate areas. A week later, the United States declared an arms embargo to the belligerents. In reality, only one side was severely affected by the embargo—the Jews. The Arab side was protected by surrounding Arab states that sent in five armies to exterminate the Jews and the newborn state the moment it was declared on May 14, 1948.

Glenn King had died three weeks earlier in a horrific air crash. His C-46 lumbered down the runway of Mexico City, and one engine began spitting and smoking as the plane strained to get off the ground. The ship clawed at the thin air, overloaded with a cargo of weapons bound for Israel before American FBI agents could stop them. The C-46 rose and crashed in a violent, fiery orange ball of death. Glenn King, the flight engineer, was killed instantly. Bill Gerson, the pilot, died a few hours later. They were the first casualties of the Israeli Air Force. Glenn was a Christian, so ironically, a Christian was the first casualty of the Israeli Air Force. Al Schwimmer, the legendary scrounger who helped assemble the rudiments of an Air Force from thin air, flew to Mexico City to claim the bodies and bring them back to America. Glenn was 31. He left a widow and children. The book I was reading said that Glenn was buried at the end of the runway in Burbank, California. "Rather curious imagery," I pondered. I pictured three dark-suited men with shovels burying King in the black of the night. Nothing more was noted about where he rested.

I was making a connection from Spokane through Burbank to Denver and continuing on to New York with a four-hour layover in Burbank. It was an opportunity and a duty to try to find Glenn's grave and pay my respects to this forgotten hero. Perhaps even inappropriately for some, I wanted to say Kaddish at his grave. After a search of maps, several calls, emails, some detective work, and luck, I discovered there actually was a cemetery at the end of the Burbank runway: the Pierce Brothers, Valhalla Memorial Park Cemetery.

The cemetery is the resting place for many famous early aviation pioneers, including Glenn King. I don't believe it was chosen because of the famous aviators. It was chosen because the Haganah Jewish defense forces used the then-isolated and remote field to smuggle weapons out of the United States. We—myself and the Hispanic groundskeeper—found the gravesite, a simple flush-to-the-ground stone plaque. It read: Glenn E. King, "A Cobber," 1917–1931. "A Cobber"? What was a Cobber?

When I arrived in Denver, I researched the meaning, and it turned out to be an Australian World War II colloquialism meaning friend, or "mate," as the Australians say. They would say to each other, "G'day Cobber," which, of course, translates to, "Hello, friend." Curiously, it is somewhat related to the Hebrew word Chaver. They are pronounced similarly and spelled differently, but both mean, for all intents and purposes, the same thing. Whoever organized the stone for Glenn King, probably an Australian, obviously recognized him as a friend.

My final stop for the week was New York. The American Veterans of Israel invited me to come to West Point for an annual memorial service for Col. Mickey Marcus. He was a Jewish graduate of West Point who served America with duty and honor in World War II. When the struggle for Israel's life against those who wished to exterminate her became real, he volunteered to serve with the Israel Defense Forces. Marcus brilliantly reorganized and led the ramshackle, virtually raw refugees fresh from the Death Camps of Europe. He created a disciplined fighting force to stop the five Arab armies who had invaded Israel. Marcus was a brilliant tactician. One night, he went outside of his lines to reconnoiter. Not speaking Hebrew, he was stopped by a sentry who misunderstood his English answer and shot Marcus. Tragically, Marcus died just before the liberation of Jerusalem. His body was repatriated to the United States, and he was interred at West Point. The AVI has an annual service to remember Marcus. This year was no different, but on the other hand, it was very different. The service to remember Marcus was also to honor and remember the non-Jewish friends of Israel who had served in the Israel Defense Forces during the War of Independence. The Jewish volunteers came for clear enough reasons. The world had stood largely silent and impotently immobile while Hitler and his supporters murdered 6,000,000 Jews. But the Christian volunteers were more of an enigma. They were few in number, but they came just the same. They came for many different reasons. Some for adventure, some as mercenaries, some for religious reasons, but almost every one of them came because they knew it was the right thing to do.

The Holocaust was not some vaguely known horror; it was a well-known monstrosity of humanity destroying innocent humanity that they could not permit to happen again. They all knew that choosing to fight for Israel was most likely a fool's effort. But they came just the same. Some never left. Some of them rest to this day in the soil of Israel. They are honored and respected, but as time and history have marched on, so has the memory of their names. But not on this Sunday in West Point.

The AVI intended to recognize them this weekend, Sunday, May 1, 2011, in a ceremony filled with pride, honor, respect, and sorrow. May 1 turned out to be a day of many meanings and feelings. May 1 was Yom HaShoah in the States. May 1 was the day that Hitler was confirmed dead. May 1 was the day that American SEALs killed a modern Hitler—Osama Bin Laden. May 1, 2011, was also the day that Jews and Christians gathered to remember their common effort to prevent another Holocaust of the Jewish people who had returned to Israel. Old men, young Cadets, Christians and Jews, children, grandchildren, and friends assembled in the Jewish Chapel at West Point overlooking the Hudson River Valley, which was just beginning to bud green with the life of spring.

The memorial program was called to order by Rafi Marom, the AVI Director. Marom's reading of a quote from the Cadet Prayer set the tone, "Make us choose the harder right instead of the easier wrong, and never to be content with a half-truth when the whole can be won." The posting of the colors was presented by the Jewish War Veterans of America Color Guard. Immediately followed with the entire assembled singing the national Anthems of the United States and Israel. Chaplain (Major) Shmuel Felsenberg, the Jewish Chaplain of West Point, delivered the invocation.

Six memorial candles were lit by cadets from the US Military Academy and Jewish students. Israel's Consul General to New York, Ido Aharoni, spoke. The Machal—overseas volunteers—both Jews and Christians, proved to play an integral part in Israel's victory in the Independence Day War. Brothers in arms, yesterday and today, Jews, Christians, and Arabs continue to serve hand-in-hand to protect the State of Israel. It is because of this brotherhood that Israel has withstood its aggressors and continues to thrive in the threat of destruction. As it is written in Isaiah 62:6-12, "I have posted watchmen on your walls, Jerusalem; they will never be silent day or night."

The Keynote address was delivered by Chaplain (Colonel) Mike Durham of the USMA. Chaplain Durham is a protestant, but for that moment in the Chapel, we were not Christians or Jews; we were one.

The names of the fallen were read out aloud, slowly, with each name rising as a messenger of the past to a future of hope. Amongst the assembled were aging Christian veterans and families of those who had gone to help Israel in her time of near-death travail.

Augustine Labaczewski, or "Duke" as he preferred to be known, sat in the front row. He was a Polish Catholic from Philadelphia who learned to speak Yiddish better than most Jews when he worked in a Jewish bakery. After the war, his friend, Mike Pearlstein, invited him to join in smuggling desperate Holocaust survivors into Palestine. Duke served on two Haganah ships—the Hatikvah and the Trade Winds—before going ashore to join the Palmach and fight in Galilee near Tiberias. He was asked to speak, but all he could say was, "You have to continue doing the right thing, do the right thing." At a quieter moment, he explained his motivations more clearly. "I think that when you learn that six million [Jews] have been killed, how can you not go? During the fighting," he said, "The only thing I could feel was that we had to win; there was no losing there."

The brother of Canada's finest World War II fighter ace, Buzz Beurling, came to represent his brother. Buzz, some said, volunteered only for the excitement, but the reality was that he came because of his deep personal faith. Buzz died in a crash in Italy. He was buried in Haifa with full military honors. Scriptures were read, a benediction completed, and the Chapel ceremony ended as the Colors were retired. We walked out quietly to the bottom of the hill, to the West Point Military cemetery, where Jew and Christian rest side by side.

The memorial service for Col. Mickey Marcus was concluded as a lone bugler played taps, and an honor guard of Cadets crisply fired their rifles in a unifying salute.

What I have learned over these past few years is that Israel exists not because of Jews or Christians alone but together as one with the ideals and hopes of many.

—Jerry Klinger, president of the Jewish American Society for Historic Preservation.

David Daniel "Mickey" Marcus was a US Army colonel, later Israel's first general, who was a principal architect of the US military's World War II civil affairs policies, including organizing the war crimes trials in Germany and Japan. He assisted Israel during the 1948 Arab–Israeli War—one of the most well-known Israeli Machal soldiers—becoming Israel's first modern general. He was killed by friendly fire.

Marcus was portrayed by Kirk Douglas in the 1966 Hollywood movie *Cast a Giant Shadow*, which focused on his role in the Israeli War.

THE MODERN ISRAELI–ARAB WAR—
THROUGH THE POWELL LENS

Here again, let's imagine a revised history through our Powell Doctrine lens. Through this filter, the United States would likely have approached the situation with a more restrained, cautious, and clearly defined strategy, focusing on specific principles that might have led to different outcomes. For example:

1. Clear Objectives

- The United States would have established well-defined, limited objectives regarding its role in the conflict, likely focusing on facilitating specific diplomatic outcomes rather than becoming deeply entangled. For example, instead of open-ended support or military involvement, the United States might have prioritized achieving a two-state solution or securing peace agreements with milestones.
- This approach would avoid shifting goals and reduce the risk of mission creep, where the United States might become further involved in regional dynamics beyond its initial intent.

2. Overwhelming Force or Decisive Action

- The Powell Doctrine's call for overwhelming force would mean that if the United States chose to become militarily involved, it would deploy sufficient resources to achieve its goals swiftly and decisively. However, given that the doctrine emphasizes action only as a last resort, it is more likely that the United States would focus on strong diplomatic pressure rather than direct military intervention.
- This principle could also lead the United States to leverage its influence to push for strong, decisive actions from Israel and Arab states in negotiations, aiming to secure agreements that would lead to a stable, long-term peace.

3. Exit Strategy

- The Powell Doctrine insists on a clear exit strategy to prevent prolonged involvement. Applied to the Arab–Israeli conflict, this would mean that the United States would seek to establish clear conditions under which it would scale back or end its involvement, such as after achieving a peace treaty or certain stability benchmarks.
- An exit strategy could involve transitioning responsibilities to local parties or international organizations, ensuring that the US role is time-bound and does not result in indefinite military or political entanglement.

4. Public and Political Support

- The Powell Doctrine values broad public and political support to ensure that any involvement has a mandate and is sustainable. The United States would likely invest more in building consensus domestically and internationally, making its involvement more transparent and well-supported.

- By aligning its actions with public sentiment, the United States could foster greater understanding and backing for its policies, potentially leading to a more consistent and unified approach over time.

5. Potential Outcomes

- **Reduced Military Presence:** The United States might limit its military involvement, focusing instead on diplomatic efforts and economic aid as tools for stability and conflict resolution. This could prevent the perception of bias and reduce anti-American sentiment in the region.
- **Focus on Conflict Resolution:** The United States would likely adopt a more neutral stance, aiming to facilitate direct negotiations between Israel and Arab states or Palestinian authorities without appearing to heavily favor one side. This could help build credibility as a mediator.
- **Avoiding Long-Term Entanglement:** By setting clear objectives and an exit strategy, the United States might avoid becoming deeply involved in the conflict's complex and long-term aspects, reducing the risk of becoming a party to the conflict itself.

In summary, employing the Powell Doctrine in the Arab–Israeli conflict might lead to a more restrained, objective-driven, and diplomatic approach. This strategy would focus on achievable goals and minimize prolonged entanglement. It could potentially foster a more sustainable and balanced role for the United States, emphasizing stability and a clear pathway to exit once objectives are met.

THE VIETNAM WAR—DATA, PROJECTIONS, AND DEPLOYING COMBAT TROOPS

The Powell Doctrine was developed based on General Powell's experiences as a commissioned officer.

General Powell served two tours in Vietnam, during which he learned a lot about leadership. After World War II, France attempted to reassert control over its former colony of Vietnam, leading to a military conflict with the communist Viet Minh forces led by Ho Chi Minh, which ultimately resulted in the Viet Minh taking control of Hanoi and North Vietnam, marking the start of the First Indochina War.

The eight-year Indochina War began in December 1946 when the Viet Minh attacked French troops in the northern part of Vietnam. The next two stages of the conflict were the People's Republic of China and the Soviet Union's recognition of the Democratic Republic of Vietnam, which was led by Ho Chi Minh. At this point, two organizations claimed to be the legitimate government of Vietnam.

During the spring of 1950, the United States announced that it would provide both military and economic aid to Vietnam's pro-French government. In 1950, the United States had nine military advisors in Vietnam. From 1950 until the fall of Dien Bien Phu in May 1954, the French forces fought the Viet Minh. The 1954 Geneva Conference that followed produced a peace treaty and left Vietnam a divided nation, with Ho Chi Minh's communist government ruling the North from Hanoi and Ngo Dinh Diem's regime, supported by the United States, ruling the South from Saigon (later Ho Chi Minh City).

In October of 1954, President Eisenhower advised South Vietnam's president that the United States would provide aid directly to South Vietnam. The last French soldier departed Vietnam in the spring of 1956.

From 1955 through 1962, the number of military advisors gradually increased over the years; during that time frame, the US Army changed from informal analysis to conducting formal war games to try to predict the future outcome of the conflict between North and South Vietnam. Starting in 1962, the Army started conducting formal war games. The war games were called the SIGMA Political-Military Games. The first game was numbered 1-62. There were four possible scenarios considered for use in game SIGMA 1-62. However, the most important fact that came out of the games that were played for 10 years is their accuracy.

The War Games were classified as Top-Secret and were played by senior US Army Officers. The projected outcomes were very accurate. For instance, the reports of the games accurately projected anti-draft riots and never projected a win by US or Allied forces engaged in the Vietnam War.

One of the questions that has never been answered is why we continued to pursue this course of action when we knew we would lose the war. Furthermore, we knew that the American citizens would not support this war. The stated reason is that this course of action was consistent with presidents Truman, Eisenhower, Kennedy, and Johnson's Doctrines.

General Colin Powell saw first-hand the damage that our leadership's approach to the war in Vietnam caused and may be the basis for the eight rules of his doctrine. Rowan Scarborough of the *Washington Times* said: "The 1980s rule said American troops would never again enter battle without decisive force and clear objectives. In other words, no more Vietnams."

It didn't take us long to forget what we learned in Vietnam. In the first Gulf War, we followed those simple rules of the Powell Doctrine and achieved wonderful results. In the Second Gulf War, we rejected both sets of rules and ended up with a disaster. Rumsfeld refused to bring sufficient forces; he brought 30% of the forces that the Army's Chief of Staff stated were necessary to win the war and the peace. As a result, we entered a war that, for all intents and purposes, is still going on. In a later chapter, we will be exploring the Sigma War Games, which reveal deep insight and raise significant questions about the Vietnam War. These games are still used today, and the data and accuracy are more accurate than ever, but the accuracy then was sufficient to question whether the decision to engage in Vietnam should ever have been made considering the overwhelming evidence to the contrary.

THE KOREAN WAR—THROUGH THE POWELL LENS

In our continuing theme, applying the Powell Doctrine to the US military action during the Korean War (1950–1953) could have led to a very different approach, potentially altering both the course and the outcome of the conflict. The Powell Doctrine, with its emphasis on clear objectives, overwhelming force, a well-defined exit strategy, and public and political support, might have influenced the decision-making process and strategies during the war. Here's how each element could have affected the Korean conflict:

1. **Clear Objectives:** One of the core principles of the Powell Doctrine is having clear, attainable, and focused objectives. In the Korean War, the initial US objective was to repel North Korean forces from South Korea following their invasion. However, after achieving this, the objective expanded to include pushing into North Korea and attempting to reunify the peninsula under a noncommunist government. If the Powell Doctrine had been applied, the United States might

have refrained from pushing beyond the 38th parallel once South Korea was secured. Instead, it could have focused on a limited objective of restoring the status quo, potentially avoiding the subsequent Chinese intervention that resulted in a costly stalemate.

2. **Overwhelming Force:** Powell advocated for using decisive and overwhelming force to accomplish military objectives quickly. After World War II, US military forces were drastically reduced. At the end of World War II, there were approximately 12 million individuals on active duty in military service. By 1947, that number had dropped to 1,566,000, a reduction of 87%. While initially, the North Koreans had vastly more forces available to them, the United States quickly recalled active-duty reserve enlisted officers. As a result of the limited number of members of the active military, the United States initially responded to the North Korean invasion with a limited force, which was pushed back by North Korean troops. If overwhelming force had been applied from the outset, the United States might have been able to stabilize the situation more rapidly, reducing the duration and intensity of the conflict. Additionally, the doctrine's emphasis on overwhelming force could have altered the decision to advance into North Korea by ensuring there were sufficient resources to handle potential Chinese intervention or by discouraging that advance altogether in favor of a contained engagement.

3. **Exit Strategy:** A well-defined exit strategy is central to the Powell Doctrine, which aims to prevent open-ended commitments. The Korean War lacked a clear exit strategy, leading to a prolonged and costly conflict that ultimately ended in a ceasefire rather than a decisive victory. Once the United States had a reasonable fighting force, the military

pushed toward the Korean/Chinese border. As a result of this activity, the Chinese entered the war and pushed the US military back, eliminating most of the United States' gains.

4. **Endgame:** If the Powell Doctrine had been in place, the United States might have planned for a more specific endgame, such as reinstating the 38th parallel boundary and rapidly withdrawing after securing South Korea. This could have minimized US involvement, avoided the costly three-year engagement, and established a more sustainable outcome.

5. **Public and Political Support:** The Powell Doctrine also underscores the importance of broad public and political support before committing to military action. The Korean War initially received public support, but as the war dragged on without a clear path to victory, public support waned. By applying the Powell Doctrine, the United States might have maintained a more focused mission that could be more easily explained and justified to the public, potentially preserving support for a shorter and more defined intervention.

Applying the Powell Doctrine might have resulted in a more restrained approach, with a focus on the limited objective of repelling North Korean forces and maintaining the boundary at the 38th parallel. This could have avoided the escalation with China, reduced US casualties, and led to a more defined and sustainable outcome. Additionally, by limiting the scope of the mission and securing public support through clear objectives and an exit strategy, the United States might have concluded its involvement in Korea sooner, potentially reducing the broader geopolitical ramifications and leading to a less volatile regional situation.

THE SOMALI WAR—THROUGH THE POWELL LENS

If the Powell Doctrine had been fully applied to the US military action in Somalia, the outcome might have been significantly different. The Powell Doctrine emphasizes the need for clear objectives, overwhelming force, a well-defined exit strategy, and public and political support. Here's how applying these principles might have altered the US mission in Somalia:

1. **Clear Objectives:** The Powell Doctrine stresses the importance of clear and attainable objectives before engaging militarily. In Somalia, the initial objective was to provide humanitarian aid and stabilize the region. However, the mission later shifted toward nation-building and attempting to capture key warlords, such as Mohamed Farrah Aidid. Had the Powell Doctrine been applied, there might have been a more focused approach to the initial humanitarian goals, avoiding mission creep into more complex and potentially unachievable political objectives.

2. **Overwhelming Force:** Powell advocated for deploying an overwhelming level of force to achieve military objectives quickly and decisively. In Somalia, US forces were not deployed in large numbers and often faced significant resistance with limited support. An example of the limited support was using light Infantry without any armor, armor-plated transport, or weapons systems. Applying the Powell Doctrine might have meant a much larger initial deployment, potentially discouraging hostilities and securing objectives with less risk to US troops. A larger force presence with appropriate equipment could have helped secure supply routes, provided robust security, and discouraged militia attacks.

3. **Exit Strategy:** The Powell Doctrine requires a clear plan for how and when to leave to avoid indefinite engagements. In Somalia, the mission lacked a clearly defined exit strategy. If the Powell Doctrine had been followed, the United States might have set a specific timeframe for humanitarian assistance and stabilization, with a clear transition plan to hand over responsibilities to local or international forces. This could have reduced the likelihood of US forces becoming entangled in prolonged conflict with local militias.

4. **Public and Political Support:** Powell also emphasized the importance of broad public and political support before military action. The humanitarian mission in Somalia initially enjoyed some public support, but as casualties mounted and the mission objectives became less clear, support waned. If the Powell Doctrine had been applied, there would likely have been a greater effort to maintain political and public backing by communicating clear, achievable goals and ensuring a rapid, effective deployment.

Applying the Powell Doctrine might have resulted in a shorter, more narrowly focused intervention in Somalia, aimed at quickly achieving specific humanitarian goals and minimizing US involvement in internal political struggles. This approach could have prevented the mission from evolving into a protracted and costly engagement, which ultimately led to US withdrawal after the "Black Hawk Down" incident. The use of overwhelming force combined with a clear exit strategy may have led to a more controlled and perhaps more successful outcome, with fewer casualties and a more limited commitment.

LIMITING THE USE OF CONVENTIONAL MILITARY WEAPONS

At the heart of warfare lies doctrine. It represents the central beliefs for waging war to achieve victory. Doctrine is of the mind, a network of faith and knowledge reinforced by experience, which lays the pattern for the utilization of men, equipment, and tactics. It is the building material for strategy. It is fundamental to sound judgment.
                                   —General Curtis E. LeMay, USAF, 1968

Why Were the Troops Unprepared for What They Encountered in Iraq?

In the second year of the occupation, the troops that arrived were told they were coming for a peacekeeping mission. As a result, they did not take most of their heavy armor. When the new units arrived, they were replacing units that either had left or were in the process of leaving. As a result, the handover of the respective area was not effective. There were very few if any, introductions of the tribal leaders with the Army's new leaders. There was no basis for trust between the locals and the Army at the company level, and very little information was transmitted at the institutional level. Therefore, the troops were unprepared.

After focused consideration by senior military officers, it was concluded that our tanks were not ideal in urban environments. According to the senior officers, the tanks and Bradley Fighting Vehicles were too imposing, and they scared civilians. Then, as calmly as he could, a Company Commander in the 2nd Battalion, 5th Cavalry Regiment, 1st Brigade Combat Team, 1st Cavalry Division Armor told his units they would be going to Iraq without their normal amount of Armor.

Records show the military had great difficulties shielding troops and their vehicles in Iraq. Initially, most of the vehicles were not armored. Even once they were up-armored, the armor was not effective against improvised explosive devices (IEDs).

Once the handover started to take place, it was almost immediately evident that the insurgency had found its stride. On a mission to clean sewage and trash, a US platoon was violently attacked, and the soldiers struggled to survive when their routine patrol of a Baghdad slum went terribly wrong.

The causalities of "Black Sunday" are depressing enough: eight soldiers died, and more than 60 were wounded during the patrol and the rescue attempt. In addition, more than 500 Iraqis were killed.

Black Sunday was only the beginning of the 1st Cavalry Division's deadly tour. After its year-long deployment, 168 of its soldiers were dead, and more than 1,900 were wounded. It was an eye-opening event for the Brigade's commander, who witnessed the complete destruction of one of his companies.

The usual mix of bad decisions by commanders in the rear and a lack of intelligence on the ground helped create the perfect mess. Department of Defense directives against soldiers serving in Iraq for longer than a year meant that the experienced soldiers leaving the part of Baghdad known as Sadr City didn't have a chance to pass along their survival techniques or intelligence information to the Black Sunday soldiers replacing them.

Despite every indication that Sadr City was "a volcano ready to explode," the Army pushed forward in the belief that the soldiers were on a humanitarian mission, busy with reconstruction projects and sucking up the knee-high slop of sewage that ran through this teeming Baghdad slum of 2.5 million people. This was "a babysitting mission"; therefore, much of the heavy armor was left at home.

The operation ran into problems from the start. The enemy was not concentrated in the villages but spread across the valley's ridges, and it was better equipped—with mortars and artillery—than intelligence had predicted.

Operation Anaconda is another example of the Army's overconfidence in its superior strength relative to the enemy. The units had poor intelligence and believed they were facing a far inferior enemy. As a result, they were directed to have only a few light-to-medium mortars and only one heavy mortar with a limited supply of ammunition—a much lower total organic fire capability than would normally be assigned to a similar army unit.

The unit was given a late notification of an operation and its scale. The issues were compounded because the air component's capability to react to the situation was reduced, resulting in several complications upon the start of the ground action.

As a result, when the demand for air support suddenly increased beyond what had been anticipated, the air component struggled to meet the suddenly urgent demands of the ground component.

A March 2009 study from the National Defense University lists "25 Problems That Occurred During Operation Anaconda." Of these, the following are the most relevant to the subject of air/ground coordination:

Lesson 1. Joint forces must continue to improve efforts to create unity of command, joint command structures, forward-deployed joint staff, and joint planning processes for expeditionary operations.

Lesson 2. Joint forces must continue to improve efforts to create unity of command, joint command structures, forward-deployed joint staff, and joint planning processes for expeditionary operations.

Lesson 3. US Joint forces need improvements in conducting integrated air-ground operations in such battles. Improvements are needed in creating a common understanding of joint force employment concepts, establishing effective information networks and joint communications systems, as well as in ensuring appropriate command and control of airstrikes in support of ground force operations

Lesson 4. US forces in battle require adequate mission orders, rules of engagement (ROE), and associated fire restrictions that give clear guidance and exert proper controls while providing force commanders the authority and latitude to execute their missions.

Lesson 5. Joint forces must be fully equipped and jointly trained for impending combat operations and (to the extent possible) surprises. Multilateral operations with allies must be well construed. Joint forces must understand the implications for training, equipping, and operating forces.

Political use of conventional weapons—Iraq-bound troops confront Rumsfeld over lack of armor. This excerpt is from a *New York Times* article that can be read in full by scanning the QR code in the footer.

"CAMP BUEHRING, Kuwait, Dec. 8. In an extraordinary exchange at this remote desert camp, Defense Secretary Donald H. Rumsfeld found himself on the defensive today, fielding pointed questions from Iraq-bound troops who complained that they were being sent into combat with insufficient protection and aging equipment.[1] It was a common complaint."

Two wonderful books, *Not a Good Day to Die* and *A Long Road Home,* also explore the details of this amazing oversight.

---

[1]   Scan code for New York Times article.

FIRST GULF GROUND WAR VS. SECOND GULF GROUND WAR

First Gulf Ground War—Powell Doctrine Led.

February 24, 1991
US and allies began a ground offensive, crossing into Iraq and Kuwait around 4:00 AM.

February 25, 1991
A Scud missile fired from Iraq destroys a barracks used by US troops in Khobar, Saudi Arabia. Twenty-eight Americans are killed, and about 100 are wounded.

February 26, 1991
Iraqi President Saddam Hussein announced that Iraq would withdraw from Kuwait immediately but did not renounce Iraq's claim to Kuwait. US and Allied forces bomb a convoy of retreating Iraqi troops, killing hundreds.

February 27, 1991
US and Allied forces entered Kuwait City. US President George Bush declares the suspension of offensive combat operations against Iraq.

Combat KIA Casualties: 145. Wounded: 467.

Second Gulf Ground War—The Shock and Awe Doctrine

- March 19, 2003: US and coalition forces invade Iraq following intelligence that the country and its dictator, Saddam Hussein, possessed or were developing WMDs.
- May 1, 2003: Bush delivers a speech aboard the aircraft carrier USS *Abraham Lincoln* proclaiming, "Mission Accomplished," saying that major combat efforts for the war in Iraq will end. "The battle of Iraq is one victory in a war on terror that began on September 11th, 2001, and still goes on," he says.
- August 19, 2003: Twenty-three people, including a top UN official, are killed and 100 wounded after a suicide bomber drives a truck into UN headquarters in Baghdad.
- December 13, 2003: Saddam Hussein is captured by US soldiers in ad-Dawr, Iraq.
- March 11, 2004: A coordinated bombing of four commuter trains in Madrid kills 191 people and injures more than 2000. Islamic militants, based in Spain but inspired by al-Qaeda, are later considered the prime suspects.
- July 7, 2005: Terrorist bombings on the London Underground and atop a double-decker bus killed 52 people and injured more than 700. Documents recovered in 2012 will reveal the attacks were planned by a British citizen working for al-Qaeda.

- December 30, 2006: After being sentenced to death by hanging for war crimes and crimes against humanity, Saddam Hussein is executed in Baghdad.
- August 30, 2010: In an Oval Office address, President Barack Obama declares an end to US combat operations in Iraq.
- May 2, 2011: Osama bin Laden is killed by US special operations forces during a raid at an Abbottabad, Pakistan, compound.
- June 22, 2011: In a televised address, Obama announces the withdrawal of US troops from Afghanistan and handover of power to Afghani security by 2014.
- August 2011: Thirty-eight service members were killed when the helicopter they were aboard came under fire. This month becomes the deadliest ever for US forces in Afghanistan, with 66 fatalities.
- December 28, 2014: The War in Afghanistan officially ends, though Obama states 10,800 US troops will remain.
- January 28, 2019: The US and Taliban leaders work toward an agreement for the withdrawal of the 14,000 US troops who remain in Afghanistan.

Casualties: 4,500. Wounded: 32,000.

Military strategists Harlan K. Ullman and James P. Wade developed the shock and awe doctrine. They introduced the concept in their 1996 book *Shock and Awe: Achieving Rapid Dominance*. The doctrine focuses on overwhelming an adversary's will to resist through a rapid and intense display of power, aiming to incapacitate the enemy's ability to respond effectively.

The Shock and Awe Doctrine was intended to result in a swift victory, minimize casualties, and avoid a protracted conflict. While it initially succeeded in toppling Saddam Hussein's regime, the doctrine faced criticism, as the prolonged insurgency that followed showed the limitations of relying on overwhelming force and rapid dominance in achieving long-term stability. The Powell Doctrine again, although in hindsight partially employed, would again have proved superior if employed in its entirety.

PREDICTING THE FUTURE—THE SIGMA WAR GAMES

The Sigma Series of "War Games" were Top-Secret US Department of Defense and Pentagon simulations designed to give various projections of how a War in Southeast Asia (Vietnam) would go. The war games started in 1962 when the US military only had a little over 3,000 advisors assigned to Vietnam.

The "Game" was regularly played by US Army Senior Officers, portraying the Enemy's "sides." The first Vietnam-centric game occurred in 1962 and was called 1-62. Sigma's results were stunningly accurate. However, most casualty ratios in the field were much higher than anticipated. Separate games were played several times most years, ending in 1972. Many scenarios also considered American society and its response to a so-called "Losing War."

The Sigma Games accurately projected anti-draft riots and other social disturbances that occurred several years after they were projected (Civil Rights Movement). The 1-62 projected that by 1972, the United States would have 500,000 troops in Vietnam, that civil unrest would occur in the United States, and that we would lose the war. No Sigma Report ever projected a clear "win" by the US or Allied forces engaged in the Vietnam War. And yet, the war went forward.

Other scenarios gamed indirect and direct combat with Communist China and Soviet Russia, both of whom were supplying war materials to the NVA and VC forces. These scenarios ended with a full-scale nuclear war at some point. Sigma made it very clear that a protracted war in Southeast Asia would eventually lead to nuclear combat. Sigma's very existence was kept secret until the National Freedom of Information Act of 1974.

War games parallel with reality with uncanny accuracy. Sigma predicted that a counter to increased US air power would be the bombardment of airfields. Sigma's gameplay was realistic enough that several plays would be replicated by actual events, such as the attacks on Bien Hoa Air Base in Saigon.

Prominent military historian H. R. McMaster terms the Sigma War Games results as "eerily prophetic."

The Sigma War Games Begin

Sigma I-62 War Game and Laotian Civil War

In February 1962, some members of the Joint Chiefs of Staff of the John F. Kennedy administration war-gamed the unfolding situation in Southeast Asia. The war game director noted, "It appears that Red wanted to win without a war while Blue wanted not to lose, also without a war." The conclusion drawn from Sigma I-62 was that American intervention would be unsuccessful. This was the first of the Sigma War Games. It occurred 3 years before President Johnson ordered the deployment of the 1st Cav (Airmobile) to Vietnam.

Sigma I-63 War Game

Sigma I-63 was played in the spring of 1963. It was held with senior-level officials setting policy for the Red and Blue Teams. Working-level officials were the actual players. Ambassador William H. Sullivan was a participant. His recollection is that Sigma I-63 ended in a fictional 1970 with 500,000 American troops locked in a stalemate in Vietnam and conscription riots in the United States.

Sigma I-64 War Game

B-52 bombing North Vietnam. Sigma suggested that air power would have little effect on North Vietnam's ability to wage war. Sigma I-64 was played between April 6 and 9, 1964. It was designed to test scenarios of escalation of warfare in Vietnam, including a gradually increasing bombing campaign. Despite a commitment of a projected 500,000 American troops to fight in Vietnam, the communists were deemed to have won.

A major change in the US commitment to Vietnam occurred during the summer of 1965. The first American combat troops arrived during the spring of 1965. Before that point in time, the United States had several thousand advisors, clerks, pilots, and other support personnel. On July 29, 1965, President Lyndon B. Johnson ordered the 1st Air Cavalry Division (also known as the Airmobile Division) to Vietnam. The division had approximately 15,000 troops. This deployment represented approximately 3% of the number of troops that Sigma 1-64 indicated would be deployed and would still result in a communist victory.

Sigma I-64 began on an imaginary June 15, 1964, with the capture of an American pilot. As Deputy Secretary of State Seymour Weiss critiqued Sigma, I-64: "The eventual capture of a

US airman is a high probability and would give 'hard' evidence of US involvement." Coincidentally, it turned out that US Navy pilot Charles Frederick Klusmann was shot down and captured in the Kingdom of Laos on June 6, 1964.

In Sigma II-64, it was predicted that General Nguyen Khanh would be pressured out of office on April 1, 1965. In real life, anti-Khanh riots broke out in November 1964, and he fled from his nation in February 1965.

Sigma II-64 War Game and Gulf of Tonkin Incident

President Lyndon Johnson and aides examining a model of a US position in South Vietnam. Sigma II-64 was scheduled as a follow-up to Sigma I-64. It was posed to answer three concerns of the US military:

(1) Would bombing the Democratic Republic of Vietnam hinder its support of the southern insurgency?
(2) Conversely, would the bombing help the South?
(3) Would they affect joint operations by the People's Army of Vietnam and the Viet Cong?

Overall, the game results were discouraging. It was concluded that raising the necessary American troops would require a state of national emergency within the United States. Most importantly, Sigma II-64's results undercut the basic assumption that a gradually escalating aerial campaign could lead to US victory. The actual conclusion was that bombing would stiffen the North Vietnamese' will to resist.

Sigma II-64 also predicted that the communists would parry American air power by bombarding airfields. When the real November 1, 1964, arrived, the Viet Cong shelled airfields at Danang and Bien Hoa for the first time, destroying six Martin B-57 Canberras.

The proposed introduction of American infantry on February 26, 1965, per Sigma II-64 really happened on March 8, 1965, when US Marines landed at Danang. In both the simulation and the reality, the United States aim was the defense of its air assets.

Another element of the Sigma II-64 scenario was a deadly ambush of an American battalion near Tchepone that inflicted heavy casualties. South Vietnamese troops during Operation Lam Son 719 in 1971 suffered heavy casualties near Tchepone (AKA Xépôn).

Sigma II-65 War Game

The Sigma II-65 war game's simulated results contradicted General William Westmoreland's strategy of attrition warfare as being capable of ending the war. As a result, Secretary of Defense Robert McNamara began to doubt the general's expertise.

French President Charles de Gaulle called for the United States' withdrawal from Vietnam in September 1966.

Sigma I-66 War Game

Sigma I-66 was staged in September 1966. Its focus was managing the de-escalation of the war if the communists were willing to begin negotiating instead of fighting. II-66 differed in that it was played to explore the effects of an outbreak of peace in Vietnam. It was based on the concept that the Vietnam War would dwindle away into defeat for the communists. To end the game, Ho Chi Minh made a secret offer to the United States to end hostilities. His requested quid pro quo was an end to the bombing campaign, withdrawal of US troops from the south, and free elections there. By the game's end, the Viet Cong were deemed the winners.

Sigma I-67 and II-67 War Games

Both these Sigma War Games were staged between November 27 and December 7, 1967. Their focus was on settling the war.

The Sigma War Games that were conducted by the US Department of Defense in the early 1960s did indeed predict potential difficulties and a likely unfavorable outcome in Vietnam. These simulations aimed to explore different military and political strategies and anticipate the possible consequences of escalating US involvement in Vietnam. Despite the pessimistic predictions of these war games, several key factors contributed to the decision to proceed with and intensify US engagement in the Vietnam War.

Containment Policy and the Domino Theory: During the Cold War, the United States was deeply committed to the policy of containment, aiming to prevent the spread of communism. The prevailing belief, often referred to as the domino theory, held that if one country in Southeast Asia fell to communism, neighboring countries would follow. This ideological perspective led US leaders to view Vietnam as a crucial battleground, regardless of the war games' predictions.

Political Pressures and Credibility: US leaders, including presidents John F. Kennedy and Lyndon B. Johnson, felt significant pressure to demonstrate resolve against communism. They believed that a withdrawal from Vietnam would damage US credibility and undermine alliances globally. For Johnson, there was also a fear that appearing weak on communism would hurt him politically at home.

Optimism Bias and Disregard for Simulations: Despite the findings of the Sigma War Games, many military and political leaders were optimistic that US technology, firepower, and economic resources would ultimately secure a victory. There was a tendency to downplay the results of simulations and view the war as winnable with the right commitment and adjustments.

Escalation Momentum: As US involvement deepened, there was a sense of "escalation momentum." After initial commitments of advisors and troops, further escalation seemed like the only viable option to avoid admitting defeat. Over time, the war effort grew in scale, making it increasingly difficult to reverse course without significant losses in prestige and perceived strategic power.

Underestimation of Viet Cong and North Vietnamese Resolve: US leadership underestimated the resolve of the Viet Cong and North Vietnamese forces, as well as their ability to mobilize and sustain prolonged conflict. There was also a misconception that conventional military superiority would be enough to overcome the guerilla tactics used by these forces, which turned out to be a critical misjudgment.

Ultimately, despite the warnings from the Sigma War Games, a complex combination of ideological, political, and strategic factors led the United States to proceed with and escalate its involvement in Vietnam, resulting in a protracted and costly conflict.

One can only imagine that the civil unrest at the time would have been amplified if the citizens had been made aware of the war games' predictions.

The Powell Lens
Comparing the Powell vs. McNamara Doctrines

Now, once more, let's explore the potential outcome of the Vietnam War through the lens of the Powell Doctrine.

If the Powell Doctrine had been applied to the Vietnam War, the US approach to the conflict might have been fundamentally different. The Powell Doctrine was developed by General Powell and was clearly influenced by his experiences as a combat leader during the Vietnam War. The Vietnam War had a significant impact on the senior officers who served as company and battalion commanders during that war. The lessons learned from it emphasize a cautious, strategic approach to military engagement, with a clear set of guiding principles for when and how to use military force. The main tenets of the Powell Doctrine, as applied to Vietnam, are as follows:

Clear and Attainable Objectives: Military action should have a clear and achievable goal. In Vietnam, objectives were often vague, ranging from containing communism to nation-building. If the Powell Doctrine had been in place, the United States might have set specific, measurable objectives, such as stabilizing South Vietnam within a defined timeframe or targeting particular North Vietnamese capabilities.

Overwhelming Force: The Powell Doctrine advocates for using decisive, overwhelming force to achieve quick and decisive victory. Rather than the incremental escalation that characterized US involvement in Vietnam, the Powell Doctrine might have called for a massive initial deployment to crush the opposition swiftly. This approach could have involved more troops, heavier bombing campaigns, or perhaps even the direct invasion of North Vietnam, all designed to achieve a quick victory with minimal protracted engagement.

Strong Public and Political Support: Powell emphasized the need for broad public and political support before committing to military action. The Vietnam War, however, was marked by significant domestic opposition and division. Had the Powell Doctrine been followed, US leaders might have sought a broader national consensus and, if this support was lacking, potentially reconsidered or limited the extent of involvement.

Exit Strategy: A clear exit strategy is crucial under the Powell Doctrine to avoid becoming bogged down in an endless conflict. In Vietnam, the lack of an exit strategy led to years of costly engagement. The Powell Doctrine would likely have required a well-defined plan for withdrawal once objectives were achieved, avoiding the kind of mission creep that marked the Vietnam War.

Vital National Interests Only: Powell emphasized that the United States should only engage militarily when vital national interests are at stake. If this standard had been applied, US policymakers might have questioned whether Vietnam truly posed a direct threat to US security and might have opted for limited involvement or even nonintervention.

Possible Outcomes of Applying the Powell Doctrine

More Decisive Initial Action or Nonengagement: By assessing whether Vietnam was truly vital to national security, the Powell Doctrine might have led to a decision not to intervene at all or to intervene in a more limited, targeted fashion.

Shorter Conflict, but Potentially Greater Initial Intensity: If the United States had deployed overwhelming force from the beginning, it might have resulted in a shorter, more intense conflict. However, this could have also escalated the risk of conflict with neighboring countries or even China, which might have been drawn in to defend North Vietnam.

Defined Objectives and Withdrawal: With clearer objectives and an exit strategy, the United States might have avoided the extended, ambiguous engagement. It could have focused on specific goals, such as stabilizing South Vietnam's government, and then withdrawn once those objectives were met.

In sum, the Powell Doctrine might have either led to a more intense but shorter intervention or even deterred US involvement altogether. By setting clearer parameters for military engagement, it likely would have avoided the protracted, costly nature of the Vietnam War, potentially leading to very different outcomes for both the United States and Vietnam.

Use of the Powell Doctrine in the First and Second Gulf Wars

During the First Gulf War, there were many stories illustrating how unprepared we were. However, by the time operations commenced, we had amassed so much might that Iraqi soldiers were surrendering to helicopters. In fact, they were surrendering to any foreign soldier who was part of the coalition because we had overwhelming might, which is why the war lasted only 96 hours.

Of the roughly 119 battlefield casualties who died during that war, 28 perished in a single Scud missile attack, while the rest died during combat operations.

Compare that to the number of deaths in the second Gulf War, when Rumsfeld ordered us to go in with 150,000 troops instead of the 600,000 we deployed in the First Gulf War.

If we review the eight decision-making criteria in the Powell Doctrine, we will see why our Army was so successful in the First Gulf War but not in the Second Gulf War, where the Powell Doctrine was not utilized.

1. Is a vital national security interest threatened?

Yes. Iraq would have a dominate market position in oil.

2. Do we have a clear attainable objective?

Yes. Destruction of Iraq's military.

3. Have the risks and costs been fully and frankly analyzed?

No. Clear disagreement in the number of troops necessary to win the initial engagement and the post-engagement activities of Iraq's military.

4. Have all other nonviolent policy means been fully exhausted?

No. There were several sanctions that had been imposed but were not being enforced.

5. Is there a plausible exit strategy to avoid endless entanglement?

No. Clear disagreement between the Army's Chief of Staff and the Secretary of Defense. The Chief of Staff was clearly correct. The war is the definition of an endless engagement.

6. Have the consequences of our action been fully considered?

No. Clear disagreement between the Army's Chief of Staff and the Secretary of Defense.

7. Is the action supported by the American people?

Unknown.

8. Do we have genuine broad international support?	Somewhat, based on misinformation.

General Rick Shinseki, who was the Chief of Staff of the Army, requested 500,000 troops, and due to this disagreement, he resigned. General Shinseki's point was that while we could clearly win the initial ground war with a smaller force, we would lose the peace because we wouldn't have sufficient forces to maintain it. And he was correct. The ninth paragraph of the Powell Doctrine states: "When a nation is engaging in war, every resource and tool should be used to achieve decisive force against the enemy, minimizing casualties and ending the conflict quickly by forcing the weaker force to capitulate."

BLUE ON BLUE

Mistakes in the Military Can Have Catastrophic, Rippling Effects

This chapter is a series of incidents to demonstrate the ramifications of seemingly small, insignificant decisions. The outcomes of incidences such as these are what inform frameworks to avoid them in the future. These frameworks could also be viewed as doctrines, which is why these examples are included.

This Blue-on-Blue Incident Marked the Start of the First Gulf War

January 17, 1991, was the first night of Operation Desert Storm. This tragic case of friendly fire—also known as "blue on blue" in military terminology (where blue refers to friendly forces)—resulted in the death of US soldiers by their own forces, a somber reminder of the chaos and confusion that often accompany large-scale military operations.

The First Gulf War began after Iraq—led by Saddam Hussein—invaded Kuwait in August 1990, prompting a massive international response. A coalition of nations, led by the United States and operating under a UN mandate, began preparing for military intervention to drive Iraqi forces out of Kuwait.

Following months of diplomatic efforts and the buildup of coalition forces in Saudi Arabia, the war commenced with Operation Desert Storm, the combat phase of the coalition's efforts to liberate Kuwait. On the night of January 17, 1991, the coalition launched a massive air campaign targeting Iraqi command and control, air defenses, and other strategic military assets.

On the first night of the operation, during the early hours of January 17, a group of US AH-64 Apache helicopters was dispatched on a mission to destroy Iraqi radar installations and air defense systems in the western part of Iraq. These Apache were part of a large, carefully coordinated effort to achieve air superiority by neutralizing Iraq's early warning radar capabilities.

At the same time, US Air Force F-15 Eagles were conducting air patrols in the same area to protect coalition aircraft from potential Iraqi fighter jets. The F-15s were equipped with advanced radar systems and air-to-air missiles designed to shoot down enemy aircraft.

During the mission, the Apache helicopters were flying low to avoid detection by Iraqi radar. Due to the fast pace of the operation, coordination between the various air units was extremely complex. At some point, the F-15 pilots detected the Apache on their radar but mistook them for Iraqi attack helicopters.

Despite the presence of identification systems designed to differentiate between friendly and enemy forces, the F-15 pilots misidentified the Apache as hostile targets and fired air-to-air missiles at them. In the ensuing attack, two Apache helicopters were shot down by the F-15s.

The friendly fire incident resulted in the deaths of two US Army pilots. This was one of the first significant casualties for US forces in the Gulf War, and it occurred before any real engagement with Iraqi forces had taken place.

Several Factors Contributed to this Tragic Blue-on-Blue Incident

Complexity of the Operation: The scale and speed of the operation led to difficulties in communication and coordination between air units. With hundreds of aircraft in the air simultaneously, confusion arose about the identification and positioning of friendly forces.

Fog of War: This term refers to the uncertainty and confusion that can occur in combat situations. On the first night of Desert Storm, amidst the chaos of launching the air campaign, even advanced systems designed to prevent friendly fire were not foolproof.

Misidentification: Despite having Identification Friend or Foe (IFF) systems in place, the F-15 pilots misidentified the Apache helicopters as enemy aircraft. Factors like radar cross-sections, communication breakdowns, and the Apache's low altitude may have contributed to the error.

Rules of Engagement (ROE): The ROE in place during the Gulf War allowed for aggressive action against potential threats, particularly in the early stages when the coalition was trying to establish air superiority. This posture, combined with misidentification, led to the mistaken attack.

The incident was a tragic beginning to the First Gulf War and underscored the dangers of modern, high-speed warfare where technology and communication systems, though advanced, can still fail under stress. It was one of several blue-on-blue incidents that occurred during the Gulf War, a reminder that even highly trained and well-equipped forces are vulnerable to errors.

In response to this and similar incidents, the US military and its coalition partners undertook further efforts to improve communication protocols, identification systems, and coordination among different branches and national forces to minimize the risk of friendly fire. These improvements became part of broader efforts to enhance battlefield situational awareness, including upgrades to IFF systems, better tracking of friendly units, and more stringent procedures for verifying the identity of targets before engaging.

A Parachuting Incident Involving the 82nd Airborne Division

What has become known as The Yuma Proving Ground incident is the common name given to the training accident of the 82nd Airborne Division. A safety officer accepted light winds as suitable conditions for a parachute jump but did not properly account for the unexpectedly strong winds at the landing zone. The landing zone in this case was at the Yuma Proving Ground, a US Army test facility in Arizona. The 82nd Airborne Division was conducting a training exercise there at the time. Weather, of course, is a critical variable to consider for any airplane or parachute operation. Wind speed is of utmost concern. Jumping into wind factors that are not properly considered can lead to serious injuries or even death.

The exercise included a safety officer who oversaw checking the wind conditions. He approved the jump after assessing the wind speeds at the takeoff point; they were within the safety limits for airborne operations. It was assumed that the conditions were just as good at the drop zone, where the paratroopers would be landing. But there was a critical oversight: no one was stationed at the drop zone to check the wind speeds there. Moreover, the wind conditions at the DZ were way too high. As the paratroopers left the plane and fell to the earth, they hit a wall of wind at the drop zone that was far stronger than anything they had encountered in the air. Many of the soldiers were blown sideways and down, with no control of their bodies, and no way to achieve a safe landing. The safety officer had no idea how bad the situation was going to be for the paratroopers.

The failure to keep adequate personnel at the drop zone to assess the wind conditions there was the single most important factor that contributed to this incident. The other side of the equation that was needed was to make it a balanced equation.

After the accident, the US Army undertook a comprehensive investigation. The Army's aviation safety experts were determined to discover not only what had happened but also why it had happened. Their first order of business after securing the crash site was to interview witnesses and survivors. The aviation safety personnel then meticulously examined the wreckage, piecing together the probable sequence of events that led to the accident. All this effort was in line with the gap-finding process inherent in any post-accident investigation.

What happened was a reminder of the basic risks we take when we fly in and out of an area. It also called into question the quality of the weather we give our soldiers. Even tiny errors can have massive consequences. The Airborne Division is the top paratrooper unit in the Army and one of the top units in the Army, period. Regrettably, it takes these seemingly benign decision-making incidents to put measures in place to avoid repeating them in the future.

The Military USB Hack—A Billion-Dollar Error

1998. Fort Campbell, Kentucky.

During this time, the internet was accessible to more people by the day, and personal computing was going places—both in the civilian world and within the military. But with all this tech came some enticing targets for would-be exploits. One of the most well-known incidents of early cyberwarfare took place in 1998 and would come to be called the "Military USB Hack." This signaled a coming storm of cybersecurity issues that would soon befall both the governmental and private sectors.

The narrative starts at a military base in the United States, where a group of engineers worked on a top-priority, top-secret project that involved creating the next generation of advanced weaponry systems. Access to information regarding the project was tightly controlled, and the engineers worked in a secure environment that was part of a military intranet. That environment was isolated from the internet and designed to prevent any kind of unauthorized access. Despite the controls in place, the engineers were not immune to the convenience that newer technologies offered. As the project matured, the team found it necessary to transfer files to and from the secure environment. They began using USB flash drives to make those transfers. That decision would, unbeknownst to them, set in motion the sequence of events that led to the project's secret being stolen.

In a bid to impress his superiors in late 1998, an intern accidentally introduced a compromised USB drive to a military secure network. A USB drive that he found, remarkably, lying in a parking lot. Unfortunately, this was no ordinary USB drive. It had been infected with a sophisticated malware designed to exploit the military's network vulnerabilities. The intern borrowed it from a friend, who had no idea that their drive was compromised. Even at a time when file sharing was already on the verge of being secured, this dubious act of transferring files seemed innocuous. However, after the intern plugged the drive into a secure system, the malware took over and began to replicate itself. The malware was programmed not just to collect information but also to appear harmless while doing so. It was siphoning off sensitive data—outgoing calls and emails, for example—to an unknown source.

The breach was discovered only after a systems administrator noticed unusual patterns of network traffic. This barely filtered into the conscious awareness of military leaders, who, to be kind, were probably as oblivious as any leadership group might be when something bad was happening. Investigators and cybersecurity experts were quickly brought in to determine the impact of the situation and to provide them with enough information so that panic would not ensue and remedial actions could be taken.

The hack that compromised military data had significant implications. It caused military authorities to completely rethink and reformulate their cyber defense strategy. In direct response to the incident, the Department of Defense instituted a series of significant changes that affected millions of uniformed and civilian personnel, as well as thousands of contractors, who work daily with the military in operating and securing vital networks. No longer would access to critical information systems be taken for granted. Access to external devices—that is, anything plugged into a computer, such as a mouse, keyboard, or, crucially, a flash drive—would from now on be part and parcel of an Access Control system that monitored, facilitated, and enabled only the right sort of people and devices to get onto the network.

The incident, once it was established that they didn't know how far the breach could have gone—even as far as the satellite system—cost over a billion dollars to rectify. Again, one tiny, seemingly insignificant act has cascading ramifications that can cost lives.

Cascading Effects of Small Errors in Aircraft Maintenance

The 2008 B-2 bomber crash at the Andersen Air Force Base in Guam offers a classic example of how a series of cascading maintenance errors and oversights can lead to a catastrophic aircraft accident. This concept, known as the Swiss Cheese Model of accident causation, suggests that errors at various levels can align in such a way that they result in a failure or accident. In the case of the B-2 crash, a combination of human error, maintenance oversight, and environmental conditions contributed to the accident.

On February 23, 2008, a B-2 Spirit Stealth Bomber crashed shortly after takeoff from the Andersen Air Force Base in Guam. The aircraft, valued at approximately $1.4 billion, was destroyed, though fortunately the two pilots ejected and survived. This was the first-ever crash of a B-2 bomber, which is one of the most advanced aircraft in the US Air Force fleet.

The Cascading Effects of Maintenance Errors

The root cause of the crash was traced to errors in the maintenance of pitot-static sensors, which are critical for measuring airspeed, altitude, and atmospheric pressure. These sensors were not properly covered during routine maintenance and were exposed to moisture, which led to incorrect readings during flight.

Error #1: Maintenance personnel did not adequately cover the pitot-static sensors when the aircraft was stored outdoors in a humid environment, allowing moisture to accumulate in them. The B-2 had been exposed to rain and high humidity while on the ground in Guam. With the pitot-static sensors compromised, moisture infiltrated the sensor system, leading to false data being fed into the aircraft's avionics.

Error #2: Improper Checks Before Flight: The moisture contamination was not identified during pre-flight checks. This was partly due to inadequate procedures for detecting such contamination and partly due to human oversight. When the B-2 took off, the sensors relayed incorrect information to the aircraft's flight control system, which relies on accurate sensor data to adjust control surfaces and ensure stable flight. Because the sensors were giving erroneous readings, the flight control system received data that indicated the aircraft's angle of attack and airspeed were different from reality.

Error #3: Misinterpretation of Sensor Data: The incorrect data caused the flight control system to initiate inappropriate adjustments to the aircraft's control surfaces, leading to a dangerously steep pitch-up shortly after takeoff. As the aircraft began to pitch up uncontrollably, the pilots attempted to regain control. However, the faulty sensor data compounded the situation by making the aircraft's behavior erratic and difficult to predict. The pilots were unable to correct the steep ascent, which caused the aircraft to stall.

Error #4: Overreliance on Automated Systems: The advanced automation and flight control systems in the B-2, while designed to assist pilots, became a liability in this case because they were receiving inaccurate data from the compromised sensors. The pilots were essentially fighting against the automated controls that were reacting to faulty inputs.

Within seconds of takeoff, the aircraft reached an extreme nose-up angle, causing an aerodynamic stall. The B-2 then lost lift and rapidly descended, crashing into the ground just off the runway. The crash destroyed the aircraft but, due to the pilots' ejection, did not result in loss of life.

The B-2 crash is a prime example of how small maintenance errors can have catastrophic effects when combined with system vulnerabilities and human factors. The following lessons were drawn from this incident:

- Importance of Proper Maintenance Protocols: Seemingly small oversights, such as failing to properly cover sensors, can have large, unintended consequences. This highlights the need for strict adherence to maintenance protocols and procedures, especially in high-stakes aviation environments.
- Comprehensive Pre-Flight Checks: Pre-flight checks are critical to identifying potential issues before take-off. In this case, the failure to detect moisture in the sensors allowed the issue to go unnoticed until it was too late. Thorough inspections, particularly after exposure to adverse weather conditions, are essential.
- Human Factors and System Reliance: Automated systems in aircraft are designed to assist pilots, but they can also introduce risks if they rely on faulty data. Pilots must be trained to recognize when automated systems are malfunctioning and be prepared to override them when necessary.
- Cascading Errors in Aircraft Crashes: The Swiss Cheese Model of accident causation was evident in this case, where multiple layers of defense (maintenance checks, sensor design, pilot response) failed, allowing a small error to cascade into a total system failure. This emphasizes the need for multiple, redundant safety measures to prevent single points of failure from escalating.

WAR CRIMES

The US invasion of Iraq in 2003, predicated largely on the assumption that Saddam Hussein possessed WMDs, remains one of the most controversial decisions in recent American history. This event was driven by a combination of intelligence failures, political motivations, and strategic calculations, all unfolding in the shadow of the September 11, 2001, terrorist attacks. Reflecting on these events reveals both the complexities of decision-making in times of crisis and the profound consequences of acting on faulty premises.

Intelligence Failures and Assumptions

The assumption that Iraq possessed WMDs stemmed from a series of intelligence reports that suggested Saddam Hussein was actively developing chemical, biological, and possibly nuclear weapons. These assessments, however, were later shown to be flawed. Intelligence agencies, under pressure to present compelling evidence, relied on unverified sources and cherry-picked information to support preexisting assumptions.

While there was widespread international concern about Hussein's ambitions, much of the evidence turned out to be circumstantial or outright false. The infamous "yellowcake" uranium claim, for example, was later debunked, as was the notion that Iraq had mobile chemical weapons labs. These intelligence failures highlight the dangers of confirmation bias, where information is interpreted in ways that reinforce existing beliefs.

The Role of the Bush Administration

In the wake of 9/11, the Bush administration pursued a proactive approach to national security, adopting the doctrine of preemptive action. President George W. Bush and key officials, such as Vice President Dick Cheney and Secretary of Defense Donald Rumsfeld, argued that Iraq posed an imminent threat to the United States and its allies. The administration also linked Iraq to al-Qaeda, despite a lack of credible evidence connecting Hussein to the terrorist organization responsible for 9/11. These arguments were presented to the American public and the international community as urgent, shaping the narrative that a swift military intervention was not only justified but necessary to prevent another catastrophe.

The Influence of Strategic Interests

Beyond concerns about WMDs, there were broader strategic interests at play. Iraq, with its vast oil reserves and strategic location in the Middle East, was seen by some as a key component in reshaping the region in a manner favorable to US interests. Neoconservative figures within the administration believed that removing Saddam Hussein and promoting democracy in Iraq could trigger a domino effect, leading to greater stability and alignment with Western values across the region. This vision of a transformed Middle East, however, underestimated the complexities of Iraqi society, including its deep sectarian divides and history of authoritarian rule.

The Fallout of the Invasion

In retrospect, the invasion of Iraq stands as a sobering reminder of the profound consequences of military intervention based on uncertain intelligence and assumptions about nation-building. The removal of Saddam Hussein created a power vacuum that led to years of sectarian violence, the rise of insurgent groups like al-Qaeda in Iraq, and ultimately the emergence of ISIS. The destabilization of Iraq also strained US resources, cost thousands of American and Iraqi lives, and left a lasting scar on US foreign policy credibility.

The invasion not only undermined trust in American intelligence but also fueled anti-American sentiment in the Middle East and beyond. By acting on inaccurate intelligence, the United States inadvertently set the stage for prolonged instability, both within Iraq and across the broader region. The lessons of Iraq have since become a cautionary tale, emphasizing the need for rigorous scrutiny of intelligence, a clear understanding of the risks of intervention, and a cautious approach to military action. The Iraq War has led to ongoing debates about accountability and the responsibility of policymakers to consider long-term consequences over short-term gains. It underscored the importance of questioning assumptions and maintaining a critical perspective, especially when making decisions with such far-reaching implications. Furthermore, it highlighted the risks of unilateral action without broad international support, as well as the limits of military power in achieving political objectives.

As the United States continues to navigate its role on the global stage, the Iraq experience serves as a reminder that the path to peace and security is rarely straightforward. It calls for a more thoughtful, informed, and measured approach to intervention— one that respects both the complexities of international relations and the profound impact of military decisions on the lives of millions. The legacy of the Iraq War challenges us to approach future conflicts with a deeper sense of caution, humility, and responsibility.

Now, let's again view the Iraq action through the lens of the Powell Doctrine. The Powell Doctrine was not fully employed in the Iraq War of 2003, although certain elements were considered. As we have already explored, the Powell Doctrine emphasizes a set of criteria for the use of military force, focusing on clear objectives, overwhelming force, strong public support, and a well-defined exit strategy. Overlying these principles with the events of the Iraq conflict, we see a clear pattern emerging in how America has played its cards in conflicts and how it might have been done differently. The clear disagreement between the Army's Chief of Staff who believed that the US need to send 500,000 troops in order to "win the peace" and the Secretary of Defense who thought that 150,000 troops combined with banning all former members of the Iraq Army and members of the Baath Party guaranteed the insurrection.

1. **Clear Objectives**

   The Powell Doctrine calls for well-defined, achievable objectives before committing to military action. While the stated goal of the Iraq War was to eliminate WMDs and remove Saddam Hussein from power, these objectives were based on flawed intelligence. Once no WMDs were found, the objectives shifted toward regime change and establishing a democratic government. This shifting of goals led to mission creep and a lack of clarity as the long-term vision for Iraq was not as well-defined as Powell would have advocated.

2. **Overwhelming Force**

The initial phase of the Iraq War did reflect Powell's preference for overwhelming force. The United States deployed a significant military presence and used its advanced technology to achieve a quick and decisive initial victory over Iraqi forces. The so-called "shock and awe" campaign was intended to demonstrate overwhelming power and quickly topple Hussein's regime. However, after the initial combat phase, troop levels were reduced, and the force was arguably insufficient to secure the country and maintain stability in the aftermath. This decision contributed to the subsequent insurgency and long-term instability, suggesting that while overwhelming force was applied at the outset, it was not sustained throughout.

3. **Strong Public and Political Support**

The Bush administration did initially secure a measure of public and political support for the invasion, largely based on claims about WMDs and links between Iraq and terrorism. However, this support was built on misinformation. As the war progressed and no WMDs were found, public opinion shifted, and support waned. This erosion of support reflects a failure to meet the Powell Doctrine's emphasis on ensuring sustained public and political backing based on transparent and accurate information.

4. **Exit Strategy**

One of the major critiques of the Iraq War is the absence of a clear exit strategy. The Powell Doctrine emphasizes the need for a well-defined plan for withdrawal once objectives are met, to avoid getting bogged down in prolonged conflicts. The United States did not have a concrete postwar plan for stabilizing Iraq or addressing the sectarian divisions within the country. As a result, the United States found itself in a protracted occupation, grappling with a deadly insurgency and struggling to rebuild and democratize Iraq—a task that proved far more complex than anticipated.

5. **Engaging Only When Vital National Interests Are at Stake**

While the Bush administration argued that Iraq posed a threat to US national security, the connection between Iraq and US vital interests was not as direct as the doctrine suggests should be the case for military intervention. The threat of WMDs was exaggerated, and Iraq's actual threat to the United States was indirect at best. This misalignment with Powell's emphasis on vital national interests contributed to the perception that the war was unnecessary.

While some elements of the Powell Doctrine were initially considered in the Iraq War, particularly the use of overwhelming force during the initial invasion, the overall approach diverged significantly from Powell's principles. The absence of clear, achievable objectives, a robust exit strategy, and sustained public support undermined the mission. Powell himself was skeptical of the war and its rationale, famously warning that "if you break it, you own it"—an admonition that highlighted the risks of destabilizing Iraq without a plan for the aftermath.

In sum, the Iraq War did not fully align with the Powell Doctrine, particularly in its long-term planning and strategic assessment. The war's eventual outcomes—prolonged instability, significant loss of life, and damaged US credibility—echoed many of the very pitfalls that the Powell Doctrine was designed to avoid.

In reflection, was Donald Rumsfeld guilty of war crimes? The answer is in all likelihood, yes. Part of that conclusion depends on his knowledge of the misrepresentations concerning Iraq's possession of weapons of mass destruction. That brings up the question of whether you would hold Colin Powell guilty of war crimes for his testimony at the United Nations, which got the vote of the United Nations to pursue the war in Iraq. The answer is no, and the reason is that he was lied to. He was provided false evidence, and all he did was provide the same evidence to the United Nations that he was provided by senior administration officials, including Donald Rumsfeld. What Rumsfeld appeared to have done was fabricate that there was evidence of WMDs being in Iraq. One issue is knowing which senior officials knew of or participated in the creation of false information concerning Iraq's possession of WMDs.

Legally, war crimes are defined under international law, particularly by the Geneva Conventions and the Rome Statute of the ICC. War crimes typically include acts like intentionally targeting civilians, using prohibited weapons, and committing genocide or crimes against humanity.

Colin Powell's testimony before the United Nations in February 2003 was based on intelligence provided to him. This testimony was a significant part of the Bush administration's case for invading Iraq. However, it was later revealed that the intelligence was flawed, and no WMDs were found. Powell himself expressed regret about his role in presenting this information, describing it as a "blot" on his record.

While Powell's testimony played a role in justifying the invasion, it does not fall under the legal definition of a war crime. The act of providing inaccurate or misleading information, even if it led to a war, does not meet the criteria for a war crime. War crimes are generally related to the conduct of war, such as actions taken during combat, rather than the decisions leading to war.

Of course, there are broader ethical debates about accountability and the responsibility of leaders who make decisions based on faulty intelligence. Some critics argue that Powell and other leaders involved in the decision to go to war should face consequences for their actions. These criticisms, however, are more about political and moral responsibility than about specific charges of war crimes. The ethical questions surrounding his role, though, continue to be a topic of debate.

The ICC did not investigate a case against Rumsfeld, partly because the United States is not a member of the ICC, and it is unlikely to participate in such a prosecution voluntarily.

While there is significant debate over the ethics and legality of Rumsfeld's actions, he was not legally convicted of a war crime regarding the fabrication of evidence about WMDs in Iraq. The question of guilt remains a matter of public and historical debate rather than legal adjudication.

Would his prosecution in this matter deter other leaders from fabricating evidence? I believe so. Otherwise, what did we learn?

True to form, Rumsfeld declared that the war in Iraq was over at the end of its 11th year, but the truth is, he made that up, too. At the time of this writing, we still have 2500 troops in Iraq.

SUPPORTING CRIMINAL ENTERPRISES MASQUERADING AS GOVERNMENTS

The US government, its State Department, the CIA, the US Army, and numerous other governmental agencies have a long and unbroken record of working with fascists, dictators, drug lords, and state sponsors of terrorism in every region of the world. The following is an abbreviated list of countries where the United States has supported fascists, drug lords, and terrorists.

1. Afghanistan now ranks 175th out of 177 countries in the world for corruption and 175th out of 186 in human development, and since 2004, it has produced an unprecedented 5,300 tons of opium per year. Ahmed Wali Karzai was well known as a CIA-backed drug lord.

2. Argentina. US documents declassified in 2003 detail conversations between US Secretary of State Henry Kissinger and Argentinian Foreign Minister Admiral Guzzetti in October 1976. After the military junta seized power in Argentina, Kissinger explicitly approved the junta's "dirty

war," in which it eventually killed up to 30,000—most of them young people—and stole 400 children from the families of their murdered parents.

3. Cambodia. President Nixon ordered the secret and illegal bombing of Cambodia in 1969. The US Defense Intelligence Agency provided the Khmer Rouge with satellite intelligence for more than a decade to counter the influence of the Vietnamese government.

4. Cuba. The United States supported the Batista dictatorship as it created the repressive conditions that led to the Cuban Revolution. The Batista dictatorship killed up to 20,000 of its own people.

5. El Salvador. The civil war that swept El Salvador in the 1980s was a popular uprising against a government that ruled with the utmost brutality. More than 70,000 people have been detained under the "state of exception," an emergency measure granting draconian powers to the police. These individuals were almost entirely established, trained, armed, and supervised by the CIA, US special forces, and the US School of the Americas.

The School of the Americas

The Western Hemisphere Institute for Security Cooperation (WHINSEC), formerly known as the School of the Americas, is a United States Department of Defense school located at Fort Moore in Columbus, Georgia, renamed in 2001 to the Western Hemisphere Institute for Security Cooperation.

The institute was founded in 1946; by 2000, more than 60,000 Latin American military, law enforcement, and security personnel had attended the school. The school was located in the Panama Canal Zone until its expulsion in 1984. Critics have labeled the institution as a school for dictators, torturers, and assassins.

My father served at the school in Panama for a period but declared it was antidemocratic and refused to return, citing, "You can't kill people who disagree with you and maintain a democracy."

The "School of the Americas" in Fort Benning, GA, which operated for 54 years as a training facility for Latin American military personnel, closed after years of criticism from human rights groups.

The list of graduates from the School of the Americas is a who's who of Latin American despots. Students have included Manuel Noriega and Omar Torrijos of Panama, Leopoldo Galtieri of Argentina, and Hugo Banzer Suarez of Bolivia.

Other graduates cut a swath through El Salvador during its civil war, being involved in the 1980 assassination of Archbishop Oscar Romero, the El Mozote massacre in which 900 peasants were killed, and the 1989 murders of six Jesuit priests.

As historian Gabriel Kolko observed in 1988, "The notion of an honest puppet is a contradiction Washington has failed to resolve anywhere in the world since 1945."

One such figure who attended SOA was Anastasio Somoza Debayle, the Nicaraguan leader who ultimately became president and led a repressive regime that resulted in widespread violence and suffering.

Somoza Regime: Anastasio Somoza Debayle was part of the Somoza family dynasty, which had ruled Nicaragua for over 40 years with US support. Somoza attended the School of the Americas in 1946, which provided him with military training that he later used to solidify his control over Nicaragua. Upon becoming president in 1967, Somoza continued to rule through

authoritarian means, maintaining power by using the Nicaraguan National Guard, which was also trained by the United
States.

Somoza's rule was marked by extensive human rights abuses,
suppression of political opposition, and a heavy-handed
approach toward dissent. His regime was notorious for corruption and brutality, with the National Guard regularly using force
to maintain control. Widespread dissatisfaction and resentment
toward Somoza's regime fueled opposition, eventually leading to
the Nicaraguan Civil War.

In the late 1970s, opposition to Somoza's rule intensified,
culminating in a full-scale civil war. The Sandinista National
Liberation Front (FSLN), a leftist revolutionary group, led a
popular uprising against Somoza's government. During the
conflict, the Somoza regime's forces killed an estimated 70,000
people, many of whom were civilians. The violent tactics
employed by Somoza's forces, including assassinations, torture,
and other atrocities, were largely aimed at suppressing the
Sandinista uprising.

Ultimately, the Sandinistas succeeded in overthrowing
Somoza in 1979. He fled Nicaragua and was later assassinated in
Paraguay in 1980. After his fall, the extent of the regime's abuses
became more widely known, and the role of the SOA in training
Latin American leaders who engaged in human rights violations
drew increased scrutiny.

The School of the Americas faced widespread criticism for
training individuals who later engaged in oppressive and violent
actions within their own countries, as seen in the case of Somoza.
The school's curriculum included counterinsurgency tactics,
which critics argued were used to suppress popular movements
and maintain authoritarian regimes in Latin America. This led
to calls for the SOA's closure, and in 2001, it was rebranded as

the Western Hemisphere Institute for Security Cooperation (WHINSEC), although concerns about its influence and role in US foreign policy persist.

Opium Production in Afghanistan

Opium production in Afghanistan generally increased during periods when the United States and its allies were involved in the country and decreased under the Taliban's initial control in the late 1990s and early 2000s.

1999-2001 Opium Ban: The Taliban, who controlled most of Afghanistan from 1996 until 2001, initially allowed opium production. However, in 2000, they imposed a strict ban on opium poppy cultivation, citing religious reasons. This led to a dramatic reduction in opium production by the following year.

The United Nations reported that opium production in Afghanistan dropped by more than 90% because of the ban, with the country producing about 185 metric tons in 2001, down from over 3,000 metric tons in 2000.

Under US and Allied Presence (2001–2021)

- **Post-Invasion Increase**: Following the US-led invasion in 2001 and the subsequent fall of the Taliban regime, opium production in Afghanistan surged. By 2002, production had already rebounded significantly.
- **Persistent Growth:** Despite counter-narcotics efforts by US and Afghan forces, opium production continued to rise over the next two decades. Afghanistan became the world's largest producer of opium, accounting for a significant majority of the global opium supply. In some years, production exceeded 8,000 metric tons.

- **Challenges in Control:** The US and Afghan governments faced considerable challenges in curbing opium production as it became a major economic activity in rural areas. The income from opium was often used to fund various groups, including the Taliban insurgency.

Taliban Control After 2021

- **Renewed Crackdown:** Since retaking control in 2021, the Taliban have again taken steps to limit opium production. In 2022, they announced a ban on opium poppy cultivation. As of recent reports, there has been a reduction in opium production, though the effectiveness and enforcement of the ban vary across regions.

So why does opium production increase with US control? Of course, it seems unlikely that the United States would promote or encourage opium production, but the active pursuit of the Taliban to support its religious beliefs had a powerful and positive effect on the control of its proliferation. Perhaps another example in support of less extended US involvement.

POLITICIZATION OF THE US MILITARY

At the end of the Afghan war, in the last five years, there was a young sergeant whose unit was on patrol, and they were ambushed. And the officers and senior enlisted people called for the QRF. It stands for quick reaction force, for reinforcement. They called repeatedly, and no one ever responded. This young sergeant proceeded to carry the wounded to safety and was wounded a lot of times himself in the process. He saved about twelve lives. They would have surely died if he hadn't carried them out and gotten them the help they needed. For that, he was awarded the Medal of Honor. He's a true hero. But the real question is, why didn't the quick reaction force come? Why didn't they respond?

There's a tactical operation center that this unit responded to, who they called repeatedly but never responded. In that operation, those who did die, died because the tactical operations center didn't respond. I can't be sure of the rank of the head of that TOC, but it's my understanding he would have been a colonel.

How in the world could a colonel not respond to a group of his soldiers being shot and killed?

Why wasn't it investigated? Is it part of the politicization of the US military that the investigation didn't happen?

* * *

After World War II, there was a program called denazification. If you were a Nazi, you were prohibited from working in the government unless you had gone through and passed the denazification board. However, this board conveniently overlooked virtually anyone who had worked in the Nazi space program. If you were involved in that program, we needed you in our own space program, so you were not considered a Nazi—even if you had been a party member the entire time and had committed war crimes. As a result, we brought these individuals to the United States.

When terrorist acts occurred after World War II, they were not as severe as those we later saw in Iraq. In Iraq, anyone who was a member of the Ba'ath Party was barred from participating in the government. This created significant problems because being a member of the Ba'ath Party was a prerequisite for holding any senior position in any agency. By removing all Ba'athists from the government, we essentially took away the jobs and incomes of many military personnel who still had weapons. What did they do in response? They resorted to theft and used their military training to fight us as insurgents.

This brings me back to the issue of the funding of 9/11 and why the information about who funded it has never been fully disclosed. The United States is heavily dependent on foreign oil for its daily operations. If one of the largest suppliers of oil to the United States was found to have funded 9/11, would it be in the best interest of politicians to reveal that and pursue action against them? What would happen to the price of oil if we did that? It would skyrocket, and politicians would likely lose their jobs as a result. However, if the government were honest with the American people and explained that we must act but that it would require reducing our oil consumption until we can find alternative sources, the public could understand and adapt.

Unfortunately, the senior US government does not trust the American people, and soldiers do not trust politicians. The only way to prevent war is to maintain a military so strong that it can inflict devastating damage on the enemy, deterring them from taking any hostile actions. But this never happens anymore because our politicians interfere.

RESIGNATION IN PROTEST
The Threat to Resign is What Gives Them the Power

On December 20, 2018, Defense Secretary James Mattis submitted a letter of resignation to President Donald Trump. Trump announced Mattis' resignation—on Twitter, no less—saying that Mattis would retire "with distinction" in February after leading the Department of Defense for two years.

In his resignation letter, Mattis, a former four-star Marine general, alluded to his policy differences with Trump, who had frequently clashed with U.S. allies.

"We must do everything we can to advance an international order that is most conducive to our security, prosperity, and values, and we are strengthened in this effort by the solidarity of our alliances," Mattis wrote in the letter.

"Because you have the right to have a Secretary of Defense whose views are better aligned with yours on these and other subjects, I believe it is right for me to step down from my position," Mattis said.

Trump praised Gen. Mattis publicly on accepting his resignation but as time went on and the true meaning of Mattis' letter became apparent, Trump's remarks became increasingly sour in relation to the General.

Peter D. Feaver, a professor of political science and public policy at Duke University, clearly articulates the premise: "The military has a legal, ethical, and professional obligation to resist illegal orders. It is not merely acceptable for the military to resist illegal orders; it is obligatory that they do so. If the President of the United States ordered General Dempsey to do something illegal, then Dempsey should resist the order up to the point of resigning in public protest. Every expert who writes or comments on this topic would agree with this point of fact. Contention on the matter arises when orders are legal but otherwise problematic or clearly act against the American people, put lives unnecessarily in danger, or otherwise put the country's security in question."

Gen. James Dubik (Ret) voiced his perspective on the matter thusly: "Waging war becomes unjust when the lives of citizens in military service are being wasted. Part of war's hellishness lies in this: war necessarily uses lives, and sometimes honest mistakes of omission and commission result in lives lost in battle. But when lives are wasted in avoidable ways like promulgating manifestly inept policies and strategies or conducting campaigns that have no reasonable chance of success because they are neither properly resourced nor connected to strategic aims worthy of the name — lives are not used, they are wasted. Senior political and military leaders are co-responsible for the lives of the citizens-now-soldiers they use in waging war."

Senator Tom Cotton asked General Milley why he did not resign after Mr. Biden rejected his advice to keep troops in Afghanistan.

General Milley said that military leaders were called to offer their advice to Mr. Biden in the lead-up to the president's April decision to withdraw. Those views, the general said, had not changed since November, when he recommended that Mr. Trump keep American troops in Afghanistan. But, the general added, "Decision makers are not required, in any manner, shape or form, to follow that advice." He also stated, "This country doesn't want generals figuring out what orders we're going to accept or not accept. That's not our job," the general replied. He later added, "My dad didn't get a choice to resign at Iwo Jima, and those kids there at Abbey Gate, they don't get a choice to resign," he said, referencing the American troops who were stationed at Hamid Karzai International Airport in Kabul in August. If the orders are illegal, we're in a different place. But if the orders are legal from the civilian authority, I intend to carry them out."

The following text outlines the strategic responsibilities of the Chairman of The Joint Chiefs of Staff.

"Statutory Notes and Related Subsidiaries EFFECTIVE DATE OF 2016 AMENDMENT Pub. L. 114–328, div. A, title IX, §921(b)(2), Dec. 23, 2016, 130 Stat. 2351, provided that: "The amendments made by paragraph (1) [amending this section] shall take effect on January 1, 2019, and shall apply to individuals appointed as Chairman of the Joint Chiefs of Staff on or after that date." § 153.

Chairman: functions (a) PLANNING; ADVICE; POLICY FORMULATION. — Subject to the authority, direction, and control of the President and the Secretary of Defense, the Chairman of the Joint Chiefs of Staff shall be responsible for the following:

(1) STRATEGIC DIRECTION. —Assisting the President and the Secretary in providing for the strategic direction of the armed forces.

(2) STRATEGIC AND CONTINGENCY PLANNING. — In matters relating to strategic and contingency planning—

(A) developing strategic frameworks and preparing strategic plans, as required, to guide the use and employment of military force and related activities across all geographic regions and military functions and domains and to sustain military efforts over different durations of time, as necessary.

(B) advising the Secretary on the production of the national defense strategy required by section 113(g) of this title and the national security strategy required by section 108 of the National Security Act of 1947 (50 U.S.C. 3043);

(C) preparing military analysis, options, and plans, as the Chairman considers appropriate, to recommend to the President and the Secretary.

(D) providing for the preparation and review of contingency plans which conform to policy guidance from the President and the Secretary; and

(E) preparing joint logistic and mobility plans to support national defense strategies and recommending the assignment of responsibilities to the armed forces in accordance with such plans. –

(3) GLOBAL MILITARY INTEGRATION. —In matters relating to global military strategic and operational integration—

(A) providing advice to the President and the Secretary on ongoing military operations; and

(B) advising the Secretary on the allocation and transfer of forces among geographic and functional combatant commands, as necessary, to address transregional, multidomain, and multifunctional threats.

(4) COMPREHENSIVE JOINT READINESS. —In matters relating to comprehensive joint readiness—

(A) evaluating the overall preparedness of the joint force to perform the responsibilities of that force under national defense strategies and to respond to significant contingencies worldwide.

(B) assessing the risks to United States missions, strategies, and military personnel that stem from shortfalls in military readiness across the armed forces, and developing risk mitigation options;

(C) advising the Secretary on critical deficiencies and strengths in joint force capabilities (including manpower, logistics, and mobility support) identified during the preparation and review of national defense strategies and contingency plans and assessing the effect of such deficiencies and strengths on meeting national security objectives and policy and on strategic plans.

(D) advising the Secretary on the missions and functions that are likely to require contractor or other external support to meet national security objectives and policy and strategy, and the risks associated with such support; and

(E) establishing and maintaining, after consultation with the commanders of the unified and specified combatant commands, a uniform system of evaluating the preparedness of each such command, and groups of commands collectively, to carry out missions assigned to the command or commands."

From the provisions of the US Code, it is clear that the Chairman of the Joint Chiefs of Staff is directly responsible for strategic, logistic, mobility, and contingency planning. While a substantial part of these responsibilities is to advise the president on the Chief's findings, it is clear they are his responsibilities.

There are times when the Chief and the President disagree on the methods for completing the strategic objectives there is a long-standing procedure for reaching a compromise. That procedure is the threat of a resignation.

In the case of the US withdrawal from Afghanistan, neither the Chief nor the Secretary of Defense offered to resign. The problem with Milly's defense of not delivering the threat to resign eliminated his possible power to get the president to rethink his position. If he were seriously committed to his position on such an important matter, threatening resignation brings "maximus gravitas" as it is no small thing to offer as an ultimatum. President Biden would have been far more likely to rethink his decision—which cost the lives of 13 U.S. service members and 10 Afghan civilians—if Milley had followed this honored procedure and been prepared to resign if his warnings were not heeded, the results might have been different.

As an author, I find Milley's behavior disgusting. General Mattis used his resignation letter to get the President to follow what he believed was a more prudent course of action. General Milley had no problem after the death of service members testifying before Congress that he had advised the President that he disagreed with the President's course of action. Again, if he were seriously committed to his position on such an important matter, threatening resignation would bring "maximus gravitas" as it is no small thing to offer as an ultimatum. I believe he failed in his moral duty to protect his subordinates." Leaders are models of courage, physical and moral. Moral courage is the courage to act under conditions of stress, to do what circumstances require, and to accept responsibility.[2] Simply put, General Milley failed as a leader by not offering his resignation.

After leaving his position as Chief—who is generally regarded as the top U.S. military officer and who is responsible for strategic matters—Milley called the 20-year war in Afghanistan a "strategic failure." One wonders what the results would have been if we had a stronger leader running the US Military.

---

[2]The Officer at Work: Leadership. Scan code to view article.

EXTENDED PEACEKEEPING—
MILITARY ARE NOT POLICE

Both General Colin Powell and General Eric Shinseki expressed concerns about the extended presence of US troops in foreign conflicts, emphasizing that the military is not designed to act as a police force or to stay indefinitely in a peacekeeping role.

1. **General Colin Powell:** Powell was clear that the military is not suited for prolonged peacekeeping or nation-building roles. He often emphasized that the military's primary mission is to achieve clear objectives, after which local authorities or other organizations should take over to maintain order and governance. Powell believed that an extended military presence could lead to mission creep, where the military ends up doing tasks outside its core competencies, such as policing and civil administration, which could undermine both military effectiveness and local stability.

2. **General Eric Shinseki:** Shinseki also highlighted that while the military can provide initial stabilization and security in the aftermath of a conflict, it is not designed to function as a long-term police force. He stressed the need for a significant number of troops to secure Iraq initially, but he also noted that extended deployments could strain military resources and divert focus from the military's primary objectives. Shinseki's warning implicitly underscored the importance of transitioning security responsibilities to local forces as quickly as feasible to prevent an indefinite military occupation.

Both Powell and Shinseki recognized that while the military could establish initial stability, an extended presence with troops acting as a police force was not sustainable or appropriate. They advocated for clear plans to hand over responsibilities to local authorities or international organizations as part of a broader exit strategy.

MODERN WARFARE AND THE POWELL DOCTRINE

The Powell Doctrine's application to hybrid warfare—a modern form of conflict that blends conventional warfare, irregular tactics, cyber operations, and state-sponsored terrorism—requires adaptation to these complex and evolving threats. Here's how the doctrine might apply:

Clear Objectives in a Hybrid Warfare Context

Original Principle: Military force should only be used if it serves a vital national interest and the objectives are clearly defined.

Adaptation:
In hybrid warfare, objectives must encompass not just physical military goals but also intangible ones like safeguarding critical infrastructure, countering disinformation campaigns, and deterring cyberattacks. For instance:

- **Cybersecurity Goals:** Defend and secure digital networks from state-sponsored or nonstate hackers.

- **Information Warfare:** Counter propaganda and disinformation campaigns that aim to destabilize societies or undermine trust in institutions.
- **Deterrence and Influence:** Use military and nonmilitary tools (e.g., sanctions, alliances, public diplomacy) to dissuade adversaries from engaging in hybrid tactics.

Clear objectives must also include measurable benchmarks for success, such as the restoration of critical systems after a cyberattack or the disruption of terrorist networks.

Overwhelming Force with Nontraditional Capabilities

Original Principle: Use overwhelming force to achieve decisive results and minimize prolonged conflict.

Adaptation:
In hybrid warfare, overwhelming force extends beyond traditional military power to include:

- **Cyber Offense and Defense:** Deploy cyber capabilities to neutralize enemy infrastructure (e.g., disrupting command-and-control networks, disabling ransomware operations).
- **Economic Sanctions:** Impose targeted sanctions to weaken adversaries economically and disrupt funding for terrorism or cyber operations.
- **Technological Superiority:** Leverage artificial intelligence, big data, and autonomous systems to gain an advantage in decision-making and battlefield awareness.

Decisive action in hybrid warfare may involve a combination of military strikes (e.g., targeting state sponsors of terrorism) and covert operations (e.g., cyber sabotage of adversary capabilities).

Broad Domestic and International Support

Original Principle: Engage only with the support of the American public and international allies.

Adaptation:
Hybrid warfare often targets alliances and public trust. To apply the Powell Doctrine effectively:

- **Public Resilience:** Build domestic consensus through transparency about threats and actions, ensuring the public understands the necessity of responses to hybrid aggression.
- **Coalition Building:** Strengthen alliances like NATO to create a unified front against hybrid threats, such as coordinated cyber defenses or intelligence sharing on state-sponsored terrorism.
- **Norms and Treaties:** Push for international agreements on cyber warfare rules, akin to the Geneva Conventions, to hold state actors accountable.

This ensures legitimacy and amplifies the effectiveness of a response, particularly in combating propaganda and cyber disinformation campaigns.

Proportionality and the Use of Force

Original Principle: Use force as a last resort, but ensure it is proportional and effective.

Adaptation:
In hybrid warfare, proportionality is complex because cyber-attacks and terrorism blur the lines of traditional conflict. Responses might include

- **Nonmilitary Countermeasures:** Deploy cyber counterattacks to disable adversaries without resorting to physical conflict.
- **Surgical Strikes:** Use precision military strikes against terrorist cells or facilities hosting hybrid operations (e.g., state-sponsored hacker groups).
- **Scalable Responses:** Implement tiered responses, from diplomatic measures to full military engagement, depending on the severity of the threat.

This ensures escalation is managed while maintaining control of the conflict narrative.

Clear Exit Strategy in Persistent Hybrid Threats

Original Principle: Avoid military engagements without a clear exit strategy to prevent entanglement in prolonged conflicts.

Adaptation:
Hybrid warfare is often designed to create prolonged, low-intensity conflicts (e.g., insurgencies, prolonged cyberattacks). Exit strategies should include

- **Resilient Infrastructure:** Build robust defenses to reduce future vulnerabilities, such as hardening critical infrastructure against cyber threats.
- **Capacity Building:** Strengthen partner nations' ability to counter hybrid threats independently, such as training allies in counter-terrorism or cyber defense.
- **Strategic Withdrawal:** Develop criteria for disengagement, such as achieving cyber stability or dismantling key terrorist networks, while ensuring ongoing monitoring and rapid response capabilities.

Exit strategies must balance achieving short-term objectives with maintaining long-term readiness for recurring hybrid threats.

Avoiding Overreach in Hybrid Warfare

Original Principle: Avoid unnecessary military action that lacks public support or exceeds national interests.

Adaptation:
Hybrid threats, particularly cyber warfare and terrorism, can provoke disproportionate responses that risk escalation. To avoid overreach:

- **Precision Targeting:** Focus on specific adversaries or networks rather than broad, undefined campaigns.
- **Legal and Ethical Boundaries:** Ensure cyber operations and counter-terrorism measures comply with international law to maintain legitimacy.
- **Strategic Patience:** Resist knee-jerk reactions to provocations, particularly in cyber warfare, where retaliation can lead to uncontrollable escalation.

Maintaining discipline in action ensures hybrid threats are managed without exhausting resources or public support.

Conclusion: Modernizing the Powell Doctrine
In hybrid warfare, the Powell Doctrine's principles remain relevant but require expansion into the domains of cyberspace, economic tools, and informational campaigns. Success depends on integrating traditional military strategies with cutting-edge technology, robust alliances, and public resilience. By adapting to these modern conditions, the doctrine can guide the next generation of military and political leaders in navigating the complex landscape of hybrid conflict.

OUR HISTORY OF DUTY OF LOYALTY TO OUR CONSTITUTION AND NOT A PERSON

This—often not discussed—early history of the United States is one of the cornerstones of the success of our country, and it all started with George Washington's first oath of office, which he drafted.

Before reviewing The Oath of Office for the President—whose terms are all contained in the Constitution—I believe a review of the historical context is necessary.

When our country came into being, George Washington was looked on so favorably that numerous individuals thought he would become America's Caesar. But that is not what Washington desired, and evidently, it is not what happened. He had a plan, and one of his first steps was—when he went to Congress to resign his commission and as commander and chief of the U.S. Military—he made several statements that I believe changed the history of our country.

1. He stated:

> *"...Having now finished the work assigned to me,*
> *I retire from the great theatre of action, and bid-*
> *ding an affectionate farewell to this august body*
> *under whose order I have so long acted, I here*
> *offer my commission and take my leave of all ap-*
> *pointments of public life..."*

So-saying, he drew out from his bosom his commission and delivered it to the President of Congress.

2. Thus, the commander—who had it in his power and who some thought aspired to, or should be a Caesar—came to Congress voluntarily to yield up his authority, in a conscious gesture, calculated to demonstrate before the world the subordination of the military to the civil authority.

> *"And let me conjure you, in the name of our com-*
> *mon Country, as you value your own sacred hon-*
> *or, as you respect the rights of humanity, and as*
> *you regard the Military and National character of*
> *America, to express your utmost horror and detes-*
> *tation of the Man who wishes, under any specious*
> *pretenses, to overturn the liberties of our Country,*
> *and who wickedly attempts to open the floodgates*
> *of civil discord, and deluge our rising empire in*
> *blood. By thus determining and thus acting, you*
> *will, by the dignity of your conduct, afford occa-*
> *sion for Posterity to say, when speaking of the*
> *glorious example you have exhibited to Mankind,*
> *'had this day been wanting, the world had never*
> *seen the last stage of perfection to which human*
> *nature is capable of attaining.'"*

3. As a result of George Washington's feelings, his oath contains only those terms required by the Constitution and carefully omits any duty to a person. The Oath of Office he took is:

"I do solemnly swear (or affirm) that I will faith-
fully execute the Office of President of the United
States and will, to the best of my Ability, preserve,
protect, and defend the Constitution of the United
States."

Many times, individuals criticized the Oath because it didn't mention people. I believe that was intentional. However, we should all remember that, at every moment—from the time he became commander in chief to his death—his project was to develop a self-governing nation, a constitutional republic. It is here that we see the brilliance of Washington's statesmanship, his hand on the political pulse of the nation, all the while urging, counseling, warning, bolstering, and leading his fellow patriots in their common efforts.

So today, we have an oath that anyone appointed to an office of honor or profit in the civil service or uniformed services (all commissioned officers) shall take the following oath:

I ___, do solemnly swear (or affirm) that I will
support and defend the Constitution of the United
States against all enemies, foreign and domestic;
that I will bear true faith and allegiance to the
same; that I take this obligation freely, without
any mental reservation or purpose of evasion; and
that I will well and faithfully discharge the duties
of the office on which I am about to enter. So, help
me, God.

Despite the fact that—when our country came into being—George Washington was looked on so favorably that numerous individuals thought he would become America's Caesar, he chose not to. Instead, we have a country where all in positions of power have sworn their duty to our Constitution, and thank God that they do.

I believe that many of our leaders have forgotten their oath and duty, which is not to a person, a party, an idea that they think is so important, or some other project. They must remember their duty is to the Constitution of the United States.

This is especially true in the military, which is expected to maintain political neutrality. However, it is very clear that many military officers act in a partisan manner or allow political considerations to influence their decisions. While this behavior is generally discouraged within the U.S. military, it appears that—as a result of a practice I, somewhat controversially, like to call "The Rummies" (An interview of individuals being selected as a general officer, instigated by Rumsfeld and during which your political affiliations were analyzed)—it is happening. The phenomenon of military officers making decisions based on partisan politics is "politicization of the military," which is a broader concern regarding the influence of politics on military decision-making and, in the future, the military influencing or, in the worst case, controlling our political decision-making.

As Washington warned us:

> *"...Express your utmost horror & detestation of the Man who wishes, under any specious pretenses, to overturn the liberties of our Country, and who wickedly attempts to open the floodgates of civil discord and deluge our rising empire in blood..."*

How close are we to this point today?

My concern is that, at times, it seems that both political parties are trying to overturn our liberties and that, as a result of this behavior, the U.S. military has been moving towards being a partisan political entity. Our new Secretary of Defense just testified as follows: "Politics should play no part in military matters. We are not Republicans or Democrats—we are American warriors." Let us all hope that he and his subordinates adhere to his promise.

EPILOGUE

The book's premise was clear: how can we do better? I believe I have demonstrated, in the same way as the Sigma War Games demonstrated outcomes, that the eight points of the Powell Doctrine should be the lens through which we filter every major military decision that faces our United States. The people deserve it, and if they don't have it, whose agenda is at play?

We have explored the American military's extensive and ongoing involvement in global conflicts since World War II. From Korea and Vietnam to the Gulf Wars and, most recently, the protracted engagements in Afghanistan and Iraq, the United States has been entangled in conflicts on nearly every continent. This examination has revealed not only the breadth of US military action but also the profound strategic, ethical, and humanitarian questions that arise from these interventions.

Reflecting on Doctrines and Leadership

At the heart of this book is the analysis of various US military doctrines and the decisions of those who wield power over the American military. Presidential doctrines, such as those of Truman, Kennedy, and Bush, have driven US foreign policy, each influenced by the unique geopolitical challenges of their time. While these doctrines aimed to protect national interests, they often led to unintended consequences, such as prolonged conflicts and regional instability. The Powell Doctrine, which demands careful justification before any military engagement, stands out as a model for strategic restraint and clear objectives—yet, as we've seen, it has not always been followed.

US leadership has also had to grapple with the complexities of coalition-building, the ethics of war, and the prioritization of strategic interests over humanitarian concerns. This book suggests that military leadership often bears the brunt of poor decision-making from the political realm, leading to situations where soldiers face morally ambiguous missions, inadequate resources, and unclear objectives.

The Cost of Politicization

The politicization of the US military has been another recurring theme, exposing the tension between military objectives and political agendas. The interference of politics in military strategy has often led to decisions that prioritize short-term political gains over long-term stability. This politicization can undermine military effectiveness, dilute the clarity of purpose, and ultimately, compromise the integrity of both the military and the mission at hand.

As we have seen, warriors—those who truly understand the gravity of war—are frequently those most critical of it. True warriors know that war is not a game or a means for political leverage but a tragic necessity that should only be a last resort. The voices of those on the front lines should be heeded, not sidelined, when discussing the costs and implications of war.

The Human Toll of War

Beyond doctrines and strategies, this book has attempted to emphasize the human cost of war, both for those who serve and the civilian populations affected by military actions. Each conflict carries with it stories of loss, sacrifice, and resilience. As we review the timeline of American military interventions, we are reminded of the millions of lives forever altered by decisions made in the corridors of power. War is not an abstract exercise but a reality that destroys families, communities, and futures.

Moving Forward: A Call for Accountability and Reflection

To avoid repeating the mistakes of the past, there must be a rigorous commitment to accountability and ethical reflection. This book advocates for a stricter adherence to frameworks like the Powell Doctrine, which encourages leaders to exhaust all non-military options, consider exit strategies, and weigh the full costs before engaging in war. Furthermore, we must hold accountable those who commit or enable war crimes, and we should bolster systems of justice that address the ramifications of these actions. Without such accountability, we not only fail those affected by past conflicts but also risk perpetuating cycles of violence and injustice.

We have explored historical wars in detail, but let's for a moment look at some contemporary actions through the same lens and see how Powell's framework might apply where the United States is not yet engaged in the action.

Ukraine–Russian War—Through the Powell Lens

The Powell Doctrine, with its emphasis on clear objectives, limited engagement, and decisive force, could offer several strategic benefits if adapted for the Ukraine–Russia conflict. Here's how its principles might be applied:

1. **Clear, Attainable Objectives:** The Powell Doctrine stresses the importance of well-defined and achievable goals. For the Ukraine–Russia conflict, this would mean setting realistic objectives for any international involvement, such as ensuring Ukrainian territorial integrity, deterring further Russian advances, or stabilizing regions impacted by conflict. This clarity would also avoid mission creep, where objectives expand over time, drawing countries into protracted involvement.

2. **Exhausting Non-Military Solutions First:** The Powell Doctrine advocates for exhausting diplomatic, economic, and political tools before resorting to military action. Applying this principle could reinforce the sanctions and diplomatic pressure already in place against Russia, potentially strengthening international coalitions and encouraging negotiations before further escalation. Increased sanctions, trade restrictions, and diplomatic isolation could serve as means to pressure Russia without immediate military involvement.

3. **Decisive Force When Necessary:** Should military action be required, the doctrine advises using overwhelming force to achieve quick and conclusive results. While direct NATO

intervention in Ukraine has been avoided to prevent escalation, this principle could still guide indirect military support. For instance, providing Ukraine with advanced weaponry, intelligence, and logistical support, along with clear parameters, would empower Ukraine's forces to counter Russian advances without direct involvement from NATO forces.

4. **Exit Strategy:** A core tenet of the Powell Doctrine is planning for disengagement from the outset. Supporting Ukrainian self-defense capabilities, rebuilding its economy, and fostering a resilient political structure could pave the way for Western nations to reduce involvement over time, as Ukraine becomes more self-reliant.

5. **Broad International Support:** Ensuring broad-based, multilateral support is essential in the Powell Doctrine. By continuing to engage NATO, the EU, and non-Western countries, the United States and allies can build a strong coalition that isolates Russia diplomatically and economically. Multilateral support also shares the burden of resources, making the strategy sustainable over the long term.

6. **Public Support and Transparency**: The doctrine emphasizes maintaining public support, especially when facing potential costs. Clear communication regarding the purpose and limits of engagement in Ukraine can help maintain domestic and allied support and ensure sustained involvement as needed.

If applied thoughtfully, the Powell Doctrine could provide structure to international support for Ukraine, minimizing the risks of overreach while bolstering Ukraine's position against Russian aggression.

Israel–Hamas War—Through the Powel Lens

The Powell Doctrine's principles could provide a framework for a measured response in the Israel–Hamas conflict by promoting clear objectives, limiting engagement, and ensuring a focused use of force. Here's how it might apply:

1. **Clear Objectives:** The Powell Doctrine calls for a defined and achievable objective. In this conflict, an objective could be the neutralization of Hamas' military capabilities to protect Israeli civilians and prevent future attacks. A focused objective avoids indefinite escalation and shifts the emphasis to clear goals, such as achieving security for Israeli civilians and restoring stability for Palestinians impacted by violence.

2. **Exhausting Non-Military Means:** The doctrine emphasizes exhausting diplomatic efforts before military action. While immediate military action is often necessary in direct defense, international pressure, diplomacy, and even indirect negotiations (often through intermediaries) could support longer-term resolutions. Working with allies to promote regional diplomacy could help prevent repeated cycles of violence, and economic incentives could address underlying conditions fueling tensions.

3. **Decisive Force When Necessary:** The doctrine advises using overwhelming force to end conflict quickly when military action is required. Israel could apply this by focusing military actions on Hamas' infrastructure and leadership rather than prolonged ground operations, which could reduce civilian casualties and the risk of a protracted war. Limiting force to strategic targets could achieve objectives

without deep entanglement in Gaza, minimizing extended warfare.

4. **Exit Strategy:** The Powell Doctrine stresses the importance of an exit strategy. In the Israel–Hamas context, this might include a phased reduction of hostilities in coordination with regional security measures and economic aid for Gaza. Clear guidelines for disengagement could help prevent long-term occupation and, ideally, foster conditions that discourage the resurgence of hostilities.

5. **Broad International Support:** Ensuring support from allies and regional partners could bolster Israel's position. The United States., the EU, and neighboring Arab states could collaborate on stabilizing efforts, emphasizing peace initiatives and humanitarian aid for Palestinian civilians. Broad support could also reinforce diplomatic efforts to bring Hamas to the negotiating table or pressure other influential actors.

6. **Public Support and Transparency:** Maintaining public support and transparency about objectives and limitations could help avoid a shift to indefinite conflict. For both domestic and international audiences, clarity about goals and commitment to minimizing civilian impact can uphold public support for necessary actions and strengthen alliances. If applied carefully, the Powell Doctrine could help Israel achieve its security objectives while setting a clear boundary to avoid prolonged entanglement, ultimately creating conditions for a more sustainable peace.

Is the Powel Doctrine a Silver Bullet?

No. As with all frameworks, processes, and guiding principles, we learn more over time, and improvements are implemented. Such is the reason we have amendments to our constitution. So how might we improve upon a proven success story such as we have examined throughout this book? Let's explore some possibilities.

We touched upon war crimes and the School of the Americas, but perhaps we could benefit from a more extensive examination of the ethical dilemmas inherent in military interventions. For example, the Just War Theory—a Christian-based doctrine that attempts to balance the importance of human life with the responsibility of states to defend their citizens—could be added to increase the flexibility of the more rigid Powell Doctrine.

The responsibility to protect (R2P) might include the use of drones and targeted killings to reduce the long-term consequences of military action on civilian populations.

Just because Powell's doctrine has proven to be the most effective overall, it does not mean that components, such as technological advancement in the Rumsfeld Doctrine or the Quantitative and Systems Analysis principles in McNamara's Doctrine, should not be considered augmentations. The Powell Doctrine emphasizes exhausting nonviolent policy means before resorting to military action, but it could be expanded to include more laser-targeted initiatives that restrict civilian casualties.

It is also important to consider how the Powell Doctrine must adapt to new threats such as artificial intelligence, cyberwarfare, and autonomous weapons systems. The idea of nanobots that may eventually be introduced could affect enemies and all of us in unimaginable ways.

Critics of the Powell Doctrine have argued that it can be too restrictive, potentially leading to inaction in situations where early intervention might be necessary to prevent humanitarian crises or stop the spread of dangerous ideologies. In scenarios such as these, contingencies could be added to allow for a more agile version of the Doctrine in demanding circumstances. Only with constant attention to improvement can we hope for a better and more peaceful future for all.

Final Thoughts

Warriors Hate War is not merely a chronicle of America's military history; it is a call to action for both citizens and leaders. As we look to the future, we must demand more transparent decision-making, increased oversight, and a renewed commitment to diplomacy over force. The American military will continue to play a crucial role in global affairs, but it is imperative that we remember the lessons of the past: that wars should be rare, measured, and justified with the utmost caution and care.

Ultimately, I wish this book to serve as a reminder that the path to a more peaceful world requires not only the strength to defend but also the wisdom to understand the true costs of conflict. In honoring those who have served, let us strive to create a world where warriors are no longer called upon to hate war but to work toward a lasting peace.

ABOUT THE AUTHOR

G lenn W. Sturm earned a Juris Doctor, with honors, from the Levin College of Law at the University of Florida, where he was named to the Order of the Coif. During law school, he served as a fellowship instructor of legal research and writing and as executive managing editor of the Florida International Law Journal. Mr. Sturm also served as an Adjunct Professor of Law at the University of Florida, where he taught Corporate Finance to law and MBA students. Mr. Sturm served in the US military as a commissioned officer on active duty and in the reserves for well over 30 years. Mr. Sturm practiced law for several decades and served on his firm's board and as its Corporate Chairman for most of his legal career. Mr. Sturm also had a successful business career. He founded Netzee Inc., an internet banking company that he took public in 1999 as its CEO. He also served on the corporate boards of directors, including Inter-Cept Inc. (ICPT), Goldleaf Financial Solutions Inc. (GFSI), and WebMD Inc. (WBMD).

In his other lives, Sturm is a photographic artist, a best-selling author, and a philanthropist who strives every day to make the lives of others better while continuing to navigate his way through over a decade and a half of life with cancer.

EXHIBITS

A   Major US military actions since World War II. We have included several of Israel's conflicts because of our financial and military aid to Israel. Furthermore, there is significant overlap of our two countries simultaneous involvement in major conflicts the same time.

Scan the QR code or visit glennsturm.com/the-powell-matrix to view The Powell Matrix, which denotes where Powell's Doctrine was utilized, not utilized, or ignored.

B   Sigma War Game reports
GlennSturm.com/sigma-war-games or scan the QR code.

Sigma War Games
a)   62 SIGMA 1-62
b)   63 SIGMA 1-63
c)   64 SIGMA 1-64 and II-66
d)   65 SIGMA 1-65
e)   66 SIGMA 1-66
f)   67 SIGMA 1-67 and II-67

www.ingramcontent.com/pod-product-compliance
Lightning Source LLC
Chambersburg PA
CBHW021917190326
41519CB00008B/812